Field Guide to the
Chessie Nature Trail

Field Guide to the
Chessie Nature Trail

Lisa Tracy and Jeanne Eichelberger
Editors

Rockbridge Area Conservation Council 2009

Republication of the Guide was made possible by a grant from the Gwathmey Trust. The original Guide was published with a grant from the Virginia Environmental Endowment.

MARINER
PUBLISHING
BUENA VISTA, VA

1 3 5 7 9 10 8 6 4 2

Field Guide To The Chessie Nature Trail
A path for all seasons

Library of Congress Control Number:
2009933454
Lisa Tracy and Jeanne Eichelberger
Editors
Includes Index and Bibliographical References

1. Natural History—Virginia—Chessie Nature Trail. 2. Chessie
Nature Trail (Va.)

I. Tracy, Lisa, 1945— II. Eichelberger, Jeanne,
1938— III. Title

ISBN 13: 978-0-9841128-1-4 (softcover : alk.
paper)

ISBN 10: 0-9841128-1-2

Mariner Publishing
A division of
Mariner Media, Inc.
131 West 21st St.
Buena Vista, VA 24416
Tel: 540-264-0021
http://www.marinermedia.com

Printed in the United States of America

This book is printed on acid-free paper meeting
the requirements of the American Standard for
Permanence of Paper for Printed Library Materials.

The Compass Rose and Pen are trademarks of Mariner Media, Inc.

Table of Contents

Table of Contents

Introduction

The Rockbridge Area Conservation Council is a volunteer, nonprofit organization established in 1976 to promote wise stewardship of natural and cultural resources through education, advocacy, and action in order to protect and enhance the quality of life for present and future inhabitants of Rockbridge County.

One of the Council's first projects was to obtain and preserve public access to the Chesapeake and Ohio Railroad right-of-way along the Maury River between Lexington and Buena Vista. The remnants of the railroad itself had been abandoned after Hurricane Camille destroyed the trestles supporting the rail bridge across the Maury at Jordan's Point in Lexington, as well as much of the track, in 1969. Through the efforts of the Council's Committee on Recreation and Trails, the potential of the right-of-way was brought to the attention of the C&O Railroad, the Nature Conservancy, and the Virginia Commission on Outdoor Recreation. Thereafter, the Conservancy accepted deed to the property in December 1978. The Conservancy transferred deed to the property to the VMI Foundation Inc. on May 8, 1979.

What you'll find along the Trail

The Trail in its entirety spans more than seven miles of the old railbed, from just west of and under the Route 11 highway bridge north of Lexington to Buena Vista. Hikers, walkers and joggers along the Trail can observe geologic features of the Maury River and its valley; plant life including a wide range of spring wildflowers and many native trees and other seed plants; animal life including mammals, birds and insects; and human traces ranging from prehistoric to the canal and railroad eras of the Trail.

The Trail is a public right-of-way, but it passes through private land for almost all of its stretch. The public is asked to be respectful of that adjacent private land. Please do not cross pasture land, for example, to access the river. Spots where canoes and kayaks may be launched are mentioned in the Trail log, and a number of important historic/archaeological sites may be seen from the Trail. School groups will find a great variety of plant life, geological features and evidence of the Trail area's animal life without need to trespass, and the river itself can be accessed at several points for water studies.

How to get there

The Trail can be accessed at four major points, going west to east – from the official head of the Trail in Lexington to its terminus in Buena Vista. These are: Jordan's Point Park, just southwest of the Route 11 bridge, along the river below the back of VMI; the Mill Creek gate, six-tenths of a mile down Old Buena Vista Road going east from the north end of the highway bridge; the South River parking area, which is reached by traversing Old Buena Vista Road east to Stuartsburg Pike (turn right and continue until just before the South River highway bridge); and from the east end of the Robey Bridge in Buena Vista. All of these access points have some parking, and at the time of this printing, funding was being sought for signage for those areas.

Once there, you'll notice more than one set of mileposts or mile markers. The original markers are the tall ones, and they take Jordan's Point as their zero point. So at about the Mill Creek gate, that's Mile 1 on the original Trail. Because many hikers, joggers and runners start from the Mill Creek gate and want to record their distances, a second, shorter set of markers begins there with the milepost marked "Start/Finish." It's expected that the Trail will be re-marked before the next edition of the Guide.

How the Trail was developed

The Foundation sought support from the Virginia Environmental Endowment to help develop the Trail for the public "recreational, educational and aesthetic" use set forth in the Deed of Transfer. At the time, the Foundation identified two prime objectives: the preservation of the right-of-way as a scenic and historic resource; and the judicious development of the railbed for hikers, joggers and nature enthusiasts. In February 1980, the Virginia Environmental Endowment awarded the VMI Foundation a $40,000 matching grant for the preservation and limited development of the 7.2-mile section of the railbed running from East Lexington to the outskirts of Buena Vista. When matching funds were obtained, the Foundation began its improvement program: clearing the right-of-way, constructing a footbridge over the Maury using the old concrete rail bridge footings as a basis, planking the steel railroad trestle over the South River, building a wooden footbridge over Mill Creek – where it crosses the trail down Old Buena Vista Road from the Route 11 bridge – and improving the roadbed throughout to facilitate walking and jogging. The Chessie Nature Trail was formally dedicated in November 1981 ceremonies at Jordan's Point on the Maury, where this outstanding Rockbridge scenic and cultural asset begins.

How the Field Guide was created

The trail was an instant success, attracting increasing numbers of individuals, as well as groups ranging from local school and college classes, Scouts, and nature study clubs to the Virginia Canals and Navigation Society.

To meet the growing demand for more information about the Trail and its environment, the Rockbridge Area Conservation Council undertook to prepare a field guide intended to encourage the educational use of the walkway through a series of essays covering both the natural and the man-made environments.

The future

As of this printing, the VMI Foundation was in the process of a deed transfer to Virginia Military Institute itself. VMI intends to repair several breaks in the Trail that have occurred in subsequent storm and flood events since the Trail's establishment. Updates on the restoration of the Trail will be accessible for hikers and others interested in the Trail through a Chessie link on the Rockbridge Area Conservation Council (RACC) website, http://organizations.rockbridge.net/racc.

Acknowledgments

The editors of this revised version of the *Field Guide to the Chessie Trail* gratefully acknowledge the original writers who were willing and able to update their essays for this edition. We particularly thank new contributors who took on the work of updating essays whose writers were not available to help us during this time. We also remember those original writers who are no longer with us, without whose work the long-lived original edition would never have come to be and the current edition would have been impossible.

The project was launched through the efforts of Richard R. Fletcher, Royster Lyle Jr., B. McCluer Gilliam, and William R. Stubbs. The original writers and those who have stepped in to update essays for this edition are noted below in the "Contributors" section. Without the tireless efforts of its original editor, Larry Bland, the entire project would most certainly never have come to completion.

The grant for republication of the Guide was made available through the generosity of the Gwathmey Trust. RACC wishes to thank Dr. David Ellington for suggesting it and Dr. Richard Brandt for his help and advice during the grant process. The writing of the grant would not have been possible without the help of RACC executive director Barbara Walsh and former RACC presidents Lee Merrill and Alexia Smith.

For this edition, most of the photographs have been re-shot to address the technological advances of the past 20 years. We gratefully thank our current photographers, also noted below.

The notable exceptions are the Miley photographs in the "Canal" and "Railroad" chapters, reprinted here courtesy of Sally Mann and the W&L Leyburn Library Special Collections Archives. Thanks also to archivist Lisa McCown, for her gracious assistance.

All of the drawings are the originals. These include the geological illustrations in Chapter 1, by Elizabeth Spencer, and bird drawings of the quail and the kingfisher by Maxine Foster.

The seed-plant illustrations come from two books by Alice Lounsberry (*A Guide to the Wild Flowers* and *Southern Wildflowers and Trees*) and from Nathaniel Lord Britton's *North American Trees*. The mushroom illustrations were taken from Thomas Taylor, *Student's Handbook of Mushrooms of America, Edible and Poisonous;* from Charles McIlvaine, *One Thousand American Fungi*; and from William Sturgis Thomas, *Field Book of Common Mushrooms*.

Mammal drawings are from Victor H. Calahone, *Mammals of North America*. Additonal bird illustrations are from Leon Augustus Hausman, *Field Book of Eastern Birds*. Butterfly drawings are from Roger G. Bland and H.E. James, *How to Know the Insects*; Austin H. Clark, *The Butterflies of the District of Columbia;* and J.H. and A.B. Comstock, *How to Know the Butterflies*.

The foldout map was originally designed and its production coordinated by Hugh Harvey. It was drawn by Ted Perdue, and hand-lettered by M. Christina Williams.

Special thanks to the Mariner Media staff, and especially Tracy Staton, Jordan Moore, Nathaniel Sidwell, Melanie Wills, Judy Rogers and publisher Andrew Wolfe, for seeing the project to completion with skill and grace.

--Lisa Tracy and Jeanne T. Eichelberger
EDITORS, 2009 Edition

Contributors

Again, we extend thanks from the Rockbridge Area Conservation Council and the editors to all the contributors whose review and updates of their original essays made this volume possible:

John Knapp is superintendent emeritus of Virginia Military Institute. He has served as Mayor of Lexington, dean of the faculty at Virginia Military Institute, and chair and professor in VMI's Civil Engineering department. An alumnus of the Institute, he holds M.S. and Ph.D. degrees from Johns Hopkins University and is a registered professional engineer. A retired major general in the U.S. Army, he is also a graduate of the U.S. Army Artillery and Infantry Schools, Command and General Staff College, and the Army War College. His field is hydrology and water resources, and he has done extensive research on Virginia's rivers, a topic on which he frequently lectures.

John Knox, a professor emeritus in biology at Washington and Lee University, holds a M.S. in botany from the University of Maryland and a Ph.D. in biology from Virginia Tech. His research interests are in plant systematics and biogeography. John updated the "Seeds" essay, which he and **Anne Knox** wrote for the Guide's original edition. Anne Knox holds a M.S. in botany from the University of Vermont and has done research on algal ecology in the Arctic.

Katie Letcher Lyle is a writer, teacher and native of Lexington who has numerous works of fiction and nonfiction to her credit, with recent books including *All Time Is Now* and *My Neighbors' Ghosts*. Educated at

Hollins College, Johns Hopkins University and Vanderbilt University, she says her interest in mushrooming was born of her interest in food and cooking, coupled with "an avid desire to avoid mushroom poisoning." Katie's culinary expertise and her ability to find morels are legendary.

John M. McDaniel, a former member of W&L's Sociology and Anthropology department, holds a Ph.D. from the University of Pennsylvania. He served as the advisor to the Regional Office of the Virginia Historic Landmark Commission's Research Center for Archaeology at Washington and Lee, starting in 1978 and continuing through the office's important study of historic and prehistoric sites in Rockbridge County, done in the early 1980s. Dr. McDaniel now lives outside of Bozeman, Montana.

Matthew W. Paxton Jr. was the editor and publisher for more than four decades of the *News-Gazette* in Lexington, a weekly newspaper that remains in the Paxton family. A Lexington native, he is a graduate of Washington and Lee University and the Columbia University Graduate School of Journalism. Railroads and their history have long been among his interests.

Robert O. Paxton grew up in Lexington and has studied birds as a hobby in Rockbridge County and farther afield for more than six decades. A professor emeritus of European history at Columbia University, he is well-known for his groundbreaking studies on France during World War II, including his 1972 book *Vichy France, Old Guard and New Order, 1940-1944*. A graduate of Washington and Lee, he holds a M.A. from Oxford University, where he was a Rhodes scholar, and a Ph.D. from Harvard University.

Edgar W. Spencer is professor emeritus and former chair of the Geology department at Washington and Lee

University. An Arkansas native, he was educated at W&L and Columbia University. He has studied the geology of the Appalachians for almost five decades. Ed is a longtime member of the Rockbridge Area Conservation Council's board of directors, and contributed a number of the photographs used in this edition.

We also thank contributors who graciously gave of their time to review others' original essays and update other aspects of the Guide:

Eliot Balazs reviewed Royster Lyle's "Historic Structures Along the Chessie Trail" and wrote new introductions for both that essay and John M. McDaniel's "Archaeological Sites" study. He is a builder in Rockbridge County, Va., currently enrolled in the Archeological Society of Virginia (ASV) certification program. He is an officer in the Upper James River Chapter of the ASV, and a member of the Rockbridge Historical Society and the Association for the Preservation of Virginia Antiquities.

Peggy Dyson-Cobb, who added the 2009 Field Study of alien and invasive species to the Seed Plants chapter, is a dedicated plants-woman and a member of the Virginia Native Plants Society. An Oberlin graduate in geology and English, she is also a member of the Rockbridge Area Conservation Council board of directors and is active in Rockbridge Grown.

Tom Kastner, who reviewed Pat Brady's "Canals" essay, is a retired naval officer who has lived in Rockbridge County since 1988. He is past president of the Virginia Canals and Navigation Society.

Barry Kinzie is the property owner and mentor of Woodpecker Ridge Nature Center in Botetourt County. A dedicated tracker of local and migrating birds and butterflies in the area, he shared these interests with

Royster Lyle in field study over the years. Of his update of Royster's "Butterflies" essay for this edition, he noted, " I have much respect for Royster's knowledge and study of things natural." Barry has been the recipient of the 1995 Conservation Award from the Virginia Society of Ornithology and the 2009 Conservation Award from the National Society of the Daughters of the American Revolution.

Michael Pelton, who provided a new "Mammals" chapter, is professor emeritus in Wildlife Science at the University of Tennessee. He currently lives in Augusta County, and frequently speaks locally on topics relating to large mammals.

Patte Wood is a professional writer and photographer who contributed a number of the photographs for the current edition of the Guide. She studied at UC Berkeley Extension with Ellen Salwen and Larry Sultan. Her longtime professional inspiration has been Henri Cartier-Bresson, whom she had an opportunity to meet while in Paris as a visitor in residence at IRCAM at the Centre Pompidou. Her photos are in private collections in Paris, Marseilles and Barcelona, and in the United States. The photos for the Guide were taken using a Zeiss Vario-Sonnar 2.8 lens.

And we gratefully remember the writers and the editor who are no longer with us:

D.E. "Pat" Brady, a Rockbridge County native, wrote the original "Canals" chapter. Mr. Brady was mayor of Lexington and superintendent of buildings and grounds at Washington and Lee University. His historical research included the canal system and the local iron industry, in which his family was involved.

Royster Lyle Jr. represented the Nature Conservancy

in this area during the early negotiations for the Chessie Trail. He was curator of collections at the George C. Marshall Foundation and coauthor of books including *The Architecture of Historic Lexington* with Pamela Simpson and Sally Mann. Royster studied butterflies and moths from his youth onward and was also a scholar of history and the author of numerous articles on the outdoors and on the history of the Valley of Virginia area.

John H. Reeves Jr. was chairman of the Biology Department at Virginia Military Institute, where he taught for more than three decades. His Ph.D. was in wildlife management, and related interests included the environment and archaeology.

Larry Bland, the Guide's first editor, was also editor of *The Papers of George Catlett Marshall* at the Marshall Foundation and editor for the Rockbridge Historical Society publications.

DEDICATION

This volume is dedicated once again, as it was in the beginning, to those who were instrumental in the creation of the Chessie Nature Trail, in particular:

Royster Lyle Jr.
Edgar W. Spencer
Odell McGuire
William R. Stubbs
B. McCluer Gilliam
Todd and Faye Lowry
Richard R. Fletcher
Hart Slater
John W. Knapp
Beverly M. Read

and in the current edition to the word person who was also so much more:

Larry Bland

The Setting

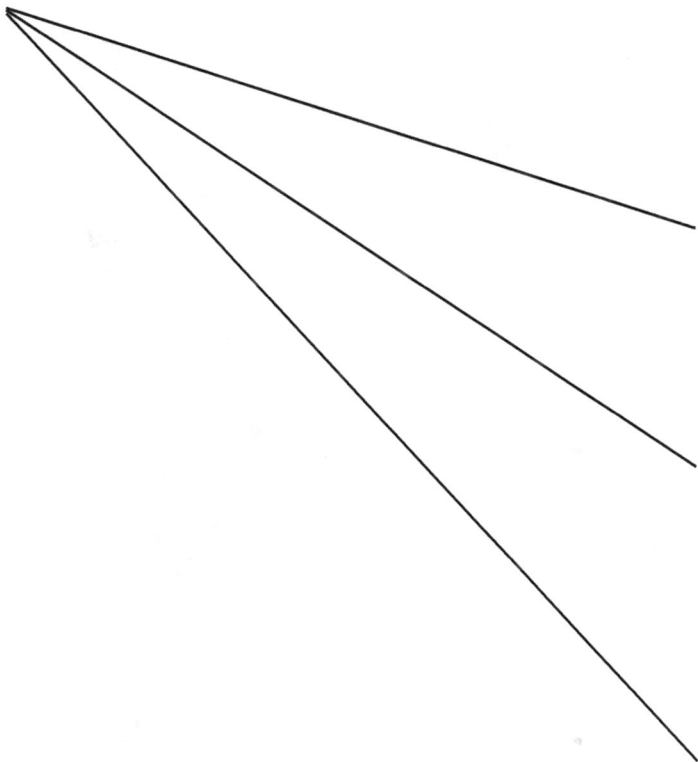

The Maury River, seen here at about the midpoint of the Chessie Trail, defines the valley through which the Trail passes.

The River and Its Valley
Edgar W. Spencer
Revised 2009

The Maury River and its tributaries provide surface drainage for most of Rockbridge County, the southwestern part of Augusta County, and a small part of Bath County, (see Fig. 1.1). The river was originally known as the North River, but was later changed to honor Matthew Fontaine Maury. Maury was the first person to map the major currents in the Atlantic Ocean, and is popularly known as "the pathfinder of the seas." He founded the U.S. Naval Observatory, but gave up his position as director at the start of the Civil War. Following the war he taught at VMI while conducting an economic survey of central western Virginia.

The name Maury is applied to the section of the river system that begins where the Calfpasture and Little

Calfpasture rivers join near Goshen Pass. It extends 38 miles to Balcony Falls, where it joins the James River close to Glasgow, Virginia. Major tributaries of the Maury include Brattons Run, Woods Creek, South River, and Buffalo Creek, Fig. 1.1. The Chessie Trail follows the section of the river that flows between Lexington and Buena Vista, on a portion of the old C&O Railroad's railbed that was destroyed by flooding in 1969.

Fig. 1.1 - Stream flow in the Rockbridge County area. The Maury River is the only river in Virginia that begins and ends in the same county.

The drainage basin of the Maury includes portions of the Great Valley of Virginia – part of the Great Appalachian Valley, as it's called, which is actually a series of valleys divided by mountains that stretches from Canada to Alabama at the heart of the Appalachian Mountains. Part of the eastern edge of the Valley and Ridge physiographic province is also in the Maury's drainage basin. Short Hills, North Mountain, and House Mountain, all of which trend northeast-southwest, are ridges of this province. These ridges – as well as the Blue Ridge, which forms the southeastern border of the

Maury drainage basin – clearly define and control the paths followed by most of the tributaries of the Maury. Like many streams in the Great Valley, most of the tributaries of the Maury flow in valleys that are underlain by easily eroded bedrock such as shale and limestone. The streams are following the course of least resistance. Between Buena Vista and Glasgow, the Maury follows a course along the northwestern edge of the Blue Ridge, as does its tributary South River. Like the bedrock in the valleys of its tributaries, the rocks beneath this section of the Maury's course are far less resistant to erosion than the rocks that compose the Blue Ridge.

Although most of its tributaries flow in valleys parallel to the intervening ridges, the Maury flows from the northwest to the southeast across the ridges at Goshen Pass and Panther Gap. It also flows across the structural grain of the underlying rock along the Chessie Trail. Shortly after the Maury joins the James, the enlarged river turns abruptly and flows southeast across the northeast-trending Blue Ridge. In addition to these paths, the Maury exhibits other surprising characteristics. Along much of its course the river flows directly on solid rock, and has cliffs on one or occasionally on both sides. The river is cutting its channel downward. In several places along its course in the Great Valley, the river has beautifully developed meanders that resemble those seen on the lower part of the Mississippi River and other rivers that are flowing across nearly flat surfaces. So the path of the Maury, as well as the sharp cliffs along its banks and the occasional meanders in its course, prompts us to wonder how this stream relates to the valley in which it flows.

The Maury River valley formed by the same processes we see in action today as we walk the Chessie Trail. The Trail Log in the back of this book points out places where you can see these processes taking place.

Although the details in the history of its formation are not clear, geologists agree that the drainage system in the Appalachians has evolved over a period of many millions of years. In this longer view, stream drainage in the Appalachians began sometime after the last major episode of mountain building, which took place toward the end of the Paleozoic era, a time that spanned the period from about 245 to 545 million years before the present (Fig. 1.2).

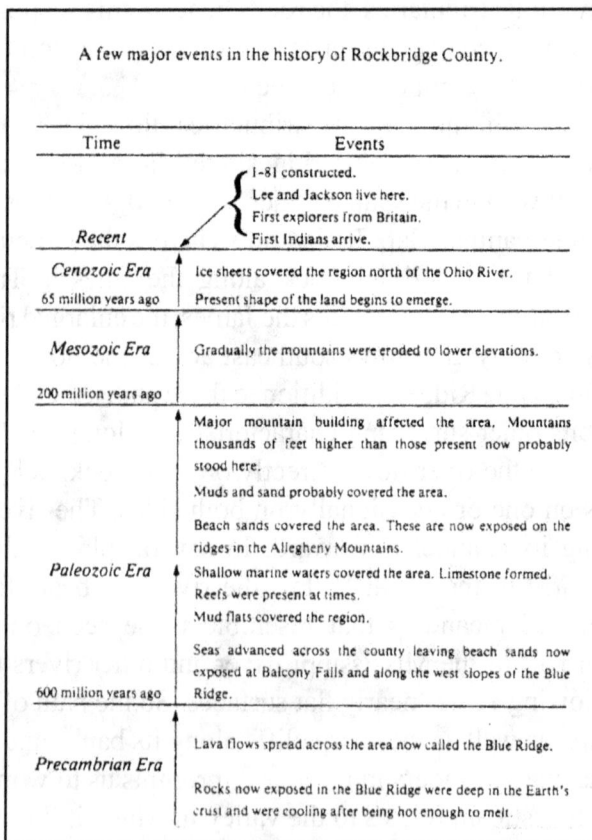

A few major events in the history of Rockbridge County.

Time	Events
Recent	I-81 constructed. Lee and Jackson live here. First explorers from Britain. First Indians arrive.
Cenozoic Era 65 million years ago	Ice sheets covered the region north of the Ohio River. Present shape of the land begins to emerge.
Mesozoic Era 200 million years ago	Gradually the mountains were eroded to lower elevations.
Paleozoic Era 600 million years ago	Major mountain building affected the area. Mountains thousands of feet higher than those present now probably stood here. Muds and sand probably covered the area. Beach sands covered the area. These are now exposed on the ridges in the Allegheny Mountains. Shallow marine waters covered the area. Limestone formed. Reefs were present at times. Mud flats covered the region. Seas advanced across the county leaving beach sands now exposed at Balcony Falls and along the west slopes of the Blue Ridge.
Precambrian Era	Lava flows spread across the area now called the Blue Ridge. Rocks now exposed in the Blue Ridge were deep in the Earth's crust and were cooling after being hot enough to melt.

Fig. 1.2

The last major mountain building in the Central Appalachians, known as the Alleghanian Orogeny, took place along the eastern margin of the North American continent when portions of the African tectonic plate collided with North America, about 300 to 350 million years ago. At that time igneous and metamorphic rocks that had formed over half a billion years earlier, along with thousands of feet of sedimentary rocks that had been deposited on top of them throughout most of the Paleozoic Era, rose to form a high mountain chain. Rocks were uplifted from depths of several miles in Earth's crust and pushed to the northwest. They were forced against, and ultimately on top of, parts of the thick pile of sedimentary rocks that had accumulated in the sea occupying the eastern margin of North America. Parts of this pile of marine Paleozoic sedimentary rocks crop out along the Chessie Trail. Everywhere along the Trail they show the effects of the uplift and crustal deformation that accompanied the mountain-building. The once-horizontal layers are tilted, some into upright positions; others are folded and turned upside down; most are inclined toward the southeast. Intense pressure from the southeast resulting from the collision of Africa with North America broke some of the layers of sedimentary rock, and those on the east side of these faults slid up and over those to the west. As a result of this deformation part of the pile is repeated as shown in Fig. 1.3. The older parts of this pile of sedimentary rocks crop out along the Blue Ridge mountain front. Somewhat younger strata lie in the Valley; still younger layers lie at the surface in the ridges such as North Mountain west of the Valley, and the youngest portions of this thick pile of sedimentary rocks crop out in the mountains of West Virginia.

Drainage of the Appalachians began to evolve on the slopes of these ancient mountains. But they were different from the mountains we know today. They probably were thousands of feet higher than the mountains we now see. When streams first began to reduce the level of the mountains through erosion, the peaks could have been as high as the modern Rockies or Alps. Streams deepen their channels and cut their valleys slowly most of the time. The debris that is currently being removed from the Maury valley can be seen in transport in the channel. It consists of gravel and sand which lie as bars on the bottom of the channel, and of the clay and silt in suspension that sometimes turn the water brown. During floods, not only this channel-fill but also material deposited earlier beside the channel as narrow flat terraces of the flood plain may be moved. Large amounts of this material were carried downstream during the floods following hurricanes Camille (1969), Agnes (1972), and Juan (1985). Enlargement and deepening of the valley is also accomplished in part as limestone dissolves and as gravel in the channel abrades and breaks up the underlying bedrock. When the banks are undercut, as they are especially on the outside of curves, the steep banks collapse into the stream. This down-slope movement of rocks and soil from valley sides can been seen at a number of places along the Trail.

As the drainage in this part of the Appalachians gradually evolved, the streams cut deeper and deeper into the uplifted mountain mass, encountering rocks of different ages, different composition and great differences in their resistance to erosion. Because the sandstones in the Valley and Ridge and the quartzites and crystalline rocks of the Blue Ridge are resistant to weathering and stream erosion, these layers of rock were etched out in the topography. In contrast, the limestones and shales exposed in the Great Valley were eroded and removed by the streams.

Thus the Great Valley is located where it is simply because it marks the outcrop belt of less resistant rocks. For the same reason, most small streams in the Appalachians flow northeast or southwest in valleys cut in this less resistant rock. South River is an excellent example of such streams. So are Woods Creek, Mill Creek, Polecat Hollow, Warm Run, Dry Branch, and Gordons Run – the small streams that flow into the Maury along the Chessie Trail. Two long stretches of the Maury, from Cedar Grove to Kerrs Creek and from Buena Vista to Glasgow, also flow along the trend of the outcrop of weak underlying rock layers. In contrast, the northwesterly trend of the Maury where it cuts through the resistant sandstones at Goshen and where the James cuts across the Blue Ridge must have some other explanation.

The geologists who first studied the landscape of this region, William Morris Davis and Douglass Johnson, concluded that the Appalachian mountains were eroded to an almost flat surface near sea level, called a peneplain. They envisioned that a marine sedimentary veneer was deposited on this surface and that it was arched upward, with the crest of the arch located just west of the present Great Valley. According to Davis, streams draining off this arch eventually formed the major southeast-flowing streams that cut across the Blue Ridge, as the James and the Potomac currently do today. Most tributaries to these major streams eventually carved courses parallel to weak layers in the bedrock. This theory has lost favor, at least in part because no remnant of the sedimentary cover can be found.

Geomorphologists who have followed Davis and Johnson also favor the view that streams cut valleys into the less resistant rocks, but they do not agree that the mountains were worn down to sea level or that the southeast course

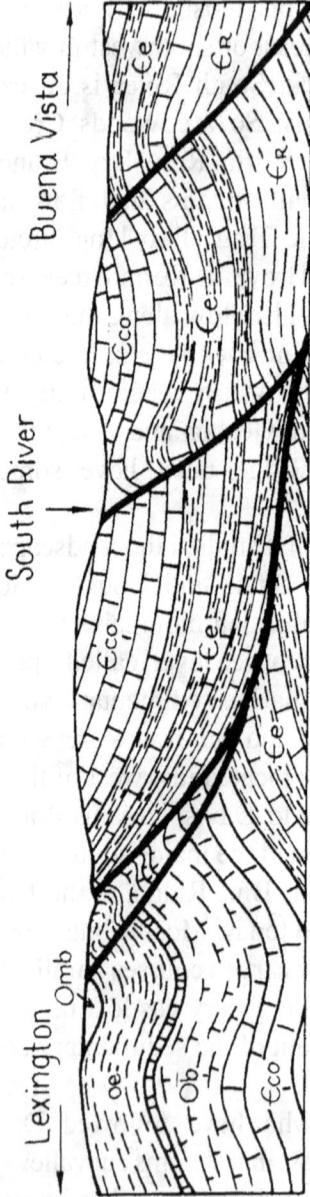

Fig 1.3 - Schematic geologic cross-section in the vicinity of the Chessie Trail.
Omb = Martinsburg formation; Oe = Edinburg formation; Ob = Beekmantown dolomite;
Cco = Conocheague formation; Ce = Elbrook formation; Cr = Rome formation

of streams like the Maury and the James were initially established on gentle southeast-sloping surfaces. Instead they envision that the Blue Ridge and the valleys and ridges to the west had already been formed before the southeast-directed drainage developed. This drainage evolved as streams, such as the James and the Potomac, flowing toward the Atlantic on the east side of the Blue Ridge, gradually extended their courses into the Blue Ridge. This took place as a result of erosion at the head of the streams where the slope of the stream channels is steep. Eventually these streams cut through the Blue Ridge, intersecting the channels of streams flowing in the valley and diverting their flow into the southeast-flowing streams. This process, known as stream piracy, is currently taking place in the Valley and Ridge Province, and is responsible for the gaps developed at Goshen and Panther Gap on the Maury River.

Both ideas about the evolution of the Maury drainage account in similar ways for the meandering path such as that seen along the Chessie Trail. Davis would have the meanders form on the peneplained surface. Recent geomorphologists would have meandering develop when the valley had low relief, and before the more recent uplift of the region caused the streams to cut their channels deeper. The meanders, originally formed on a surface of low relief, are incised, "entrenched," into solid rock. The steep cliffs along the side of the channel, the presence of abandoned channel-fills on high ground above the valley, and the exposures of bedrock in stream channels all indicate that downcutting is still in progress. Thus it seems likely that the Maury flowed in a meandering course across the valley on a relatively flat surface at about the level of the top of the cliffs now seen from the trail before it began to cut down to its present level. ❧

Photo by Patte Wood

Signs and a commemorative plaque mark the head of
the Chessie Trail in Jordan's Point Park.
During Hurricane Camille in 1969, floodwaters rose almost 13 feet
above the elevation of the base of the monument.

Hydrology of the River
John W. Knapp
Revised 2009

Hydrologists study the occurrence, the movement, and the conservation of natural waters; they consider the average yield of rivers, the pattern of high and low flows, and the potential for development and recreation. The Maury is an interesting case in applied hydrology and many examples can be seen along the Chessie Trail.

The river which runs along the Chessie Trail is more than just an important natural resource of the Rockbridge County area. Over the centuries it has physically shaped the valley it flows through, and in recent times it has contributed to the cultural and economic development of the entire region.

The vagaries of climate and weather have periodically brought damaging floods to the settlements along the river's banks, while at other times droughts have plagued the surrounding countryside. The desire to control these extremes has sometimes led to studies for damming the river, but so far only minor structures have been built, chiefly to harness the river's power for milling, manufacturing, and producing electricity. Once a major artery of commercial transportation via its mid-19th century canal system, today it is a principal source for municipal drinking water and a major recreational feature. Waste treatment facilities have been built to preserve the quality of the river's water, and land-use practices have been instituted with the hope of protecting the values of land and river for future generations.

The ordinary flow conditions in a river are determined by the size of the upstream drainage area, the amount of precipitation, and the character of the basin's rocks, soil, and vegetation. At Lexington the drainage area of the Maury is about 500 square miles. Where the river empties into the James River at Glasgow, twenty miles downstream, it drains 840 square miles. South River, near mile point 4 below Lexington, adds 92 square miles of drainage area, and Buffalo Creek between Buena Vista and Glasgow adds another 124 square miles.

The U.S. Geological Survey maintains two gauging stations on the river. One of them, the Buena Vista gauge, is below the confluence of the Maury (formerly North River) with South River, near mile point 5 of the Trail, and is easily visible on the far bank. At that point the drainage area is 649 square miles, and the 40-year average of daily stream flow is 663 cubic feet per second.

As with many streams in this area, the average flow is about one cubic foot per second for every square mile of drainage area. In one year that represents the equivalent of 14 inches spread over the entire watershed.

Since annual rainfall averages 40 inches, the difference of 26 inches between precipitation and runoff represents what is lost, in the average year, to evaporation and to transpiration by vegetation.

Of course the river seldom stays at average flow conditions. It sometimes changes by the day and always varies during the year, with flows generally increasing through the winter and early spring months and diminishing through the summer and early fall. By the end of September, the river's flow often consists entirely of ground water contributed from springs; any deficit in the ground must be made up by the next year's excess rainfall.

So it is the range of ordinary flows, the annual highs and lows, that define the principal uses to be made of the river. The high flows, roughly bank-full discharges, are primarily responsible for the visible conditions and character of the bed and banks. The annual lows define the dependable yield of the river for withdrawals, such as supplies for drinking water. Storage behind large dams could increase the dependable yield by saving water during high-flow periods for release during low-flow periods.

The flood history of a river describes those infrequent but inevitable and dramatic events caused by weather conditions that create severe storms over the basin. Sometimes the flood runoff is aggravated by icy or wet conditions, occasionally by the effects of a hurricane.

The occurrence of such events is difficult to predict and is, therefore, usually characterized by probabilities of occurrence. For example, the hundred-year flood is the flood expected every year with a chance of 1 in 100, a probability of 1 percent. It is a rare event, but one that is possible every year. Likewise, the ten-year event is expected each year with 10 percent probability. The

Maury's floods are shown in the following table:

Event	Flow
Average Daily Flow	663 c.f.s.*
2-year flood (50%)	11,500 c.f.s.
10-year flood (10%)	23,100 c.f.s.
25-year flood (4%)	31,400 c.f.s.
50-year flood (2%)	38,800 c.f.s.
100-year flood (1%)	47,300 c.f.s.

*c.f.s stands for cubic feet per second

Actual flood depths vary at every point along the river, but for comparison one might expect that the 2-year flood on the Maury would rise just above the river banks, that the 10-year flood would rise to 15 or more feet above the river bed, and that the 100-year flood would be 25 feet or more above the bed.

Maury River floods and the damage they caused to riverside industries, bridges, and farms have been observed since the late eighteenth century. Newspapers record large floods in July 1842, December 1847, April 1870, and twice in November 1877. Undoubtedly the first of the 1877 floods on November 8 left the watershed saturated for the "Great Flood" that followed on November 20, 21 and 22. This flood seems finally to have ended canal system navigation.

A flood in March 1936 was in the 50- to 100-year range and caused considerable damage in East Lexington and in Buena Vista. But the highest flood of record came in August 1969 when Hurricane Camille hit the Maury and South River basins. The estimated peak discharge at the Buena Vista gauging station was 105,000 c.f.s., which is greater than the 150-year event.

The effects were devastating in Rockbridge County, where twenty-three lives were lost. In Buena Vista damages to the low-lying sections of the city totaled millions of dollars and major disaster relief operations were required. In Lexington, the railroad trestle was carried away, leaving the concrete piers on which a footbridge at the beginning of the Chessie Trail was built. Needless to say, this event effectively ended railroad service to Lexington, and some years later the conveyance of the right of way led to the creation of the Chessie Trail.

Then in November 1985, the flood of record occurred in the upper reaches of the Maury. At Lexington the waters rose more than five feet above the 1969 level. At its peak the waters inched across Route 11 (Main Street) on the south side of the main highway bridge, near the mouth of the Route 11 bypass. In Buena Vista, however, the 1985 flood was actually a half foot lower than in 1969, mainly because the flooding on South River varied so greatly. Fortunately, no lives were lost in Rockbridge County during the 1985 flood.

The flood of June 1982, which destroyed the first Chessie Trail footbridge, was somewhere between the five-year and ten-year event in severity. The trail monument (see photograph on p. 11) was completely out of the water, although the peak waters lapped against the base. In 1969 the flood rose to a depth of 12.9 feet above the elevation of the base of the monument. In 1985 the flood was 18.1 feet above the base of the monument. In 1987, the footbridge was again over-topped and clogged with debris, and the decision was made to forgo replacing it. Today the City of Lexington is improving Jordan's Point as a public park. Plans call for a path and footbridge over Woods Creek to lead trail users over the Route 11 highway bridge to the Chessie Trail.

The low-flow characteristics of the river are reflected in the following table:

Low-Flow Event	Avg. Flow (in c.f.s.) for Consecutive Days		
	1-Day	7-Days	30-Days
2-year (50%)	85	92	107
10-year (10%)	50	59	68
100-year (1%)	42	45	55

These are considerably less than the average flow. Note that the data show the 100-year drought flows are only one-half of the low flows expected every other year (the 2-year events). These conditions imply that the river cannot be forced much lower because the carbonate rocks underlying the watershed hold large reserves of water which are released to the river during periods of drought. But even at the lowest flow shown—42 c.f.s. for one day with a probability of 1 percent occurrence—there still would be sufficient water to serve the daily domestic needs of 200,000 people. With storage behind a large dam, the number that could be served would increase significantly.

The Maury Service Authority has a water filtration plant on the river about two miles above Lexington. This plant produces drinking water for the city and for nearby areas of the county. The lowest flows in the river are more than adequate to serve eight to ten times the present demand.

For more than 50 years the City of Lexington operated a waste-water treatment plant on Jordan's Point on land that is a playing field today. In the late 1990s, the newly formed Lexington-Rockbridge Regional Water Quality Control Facility (WQCF) completed a new treatment plant downstream and northeast of Jordan's Point on Bob Akins Circle. A pumping station near mile point 1 and the mouth of Mill Creek lifts raw sewage up to the plant. The facility is owned by the Maury Service Authority and operated by the City of Lexington. Although it is designed and permitted to treat 3 million gallons per day, it is currently treating only one million gallons. It does provide what is called secondary treatment, including some removal of nutrients, primarily nitrogen. However, by 2010 the plant will be upgraded to meet increased nitrogen and phosphorus treatment standards required by a multi-state effort to improve the Chesapeake Bay.

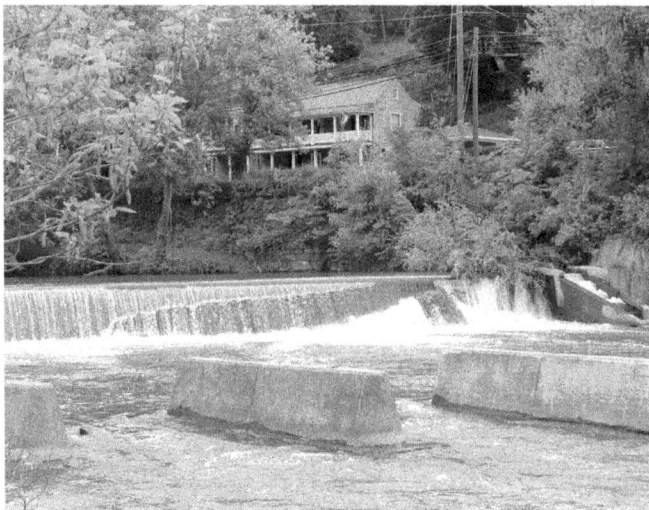

Photo by Patte Wood

The dam at Jordan's Point is just upstream from the highway bridge, as are the footings of the old railroad trestle that once supported a Chessie Trail footbridge, destroyed by flooding. In the background is a home popularly known as Old Bridge, formerly Tankersley's Tavern.

The Maury River is known as one of the best white-water canoeing streams in Virginia, but the best sections are not along the Chessie Trail. There are, however, several stretches of easy rapids below the old dam sites at Reid's, South River and Ben Salem. The best time of the year is April through June; later in the summer, a loaded canoe will scrape bottom in many places. The five- to six-mile stretch between the Trail's entrance at Jordan's Point and Ben Salem has the most convenient points of access.

During the 19th century, the power of falling water was used on the Maury for varied industrial purposes. Most of the grist mills and the iron furnaces and forges were on tributary streams. At Lexington from earliest days and later at Buena Vista waterwheels once provided the power to turn machinery.

In the early years of the 20th century at Reid's Dam, which was completely intact until Hurricane Camille caused a breach, a turbine was installed to generate electricity for Lexington. That operation ended by the 1930s, as more sophisticated and efficient power plants were built.

The river continues to serve as a valuable resource for the entire region. It is the dominant natural feature along the Chessie Nature Trail, and indeed mothered the Trail itself. It has a rich heritage, and its benefits—both tangible and intangible—must be assured for future generations by unremitting care. ❧

The Natural Environment

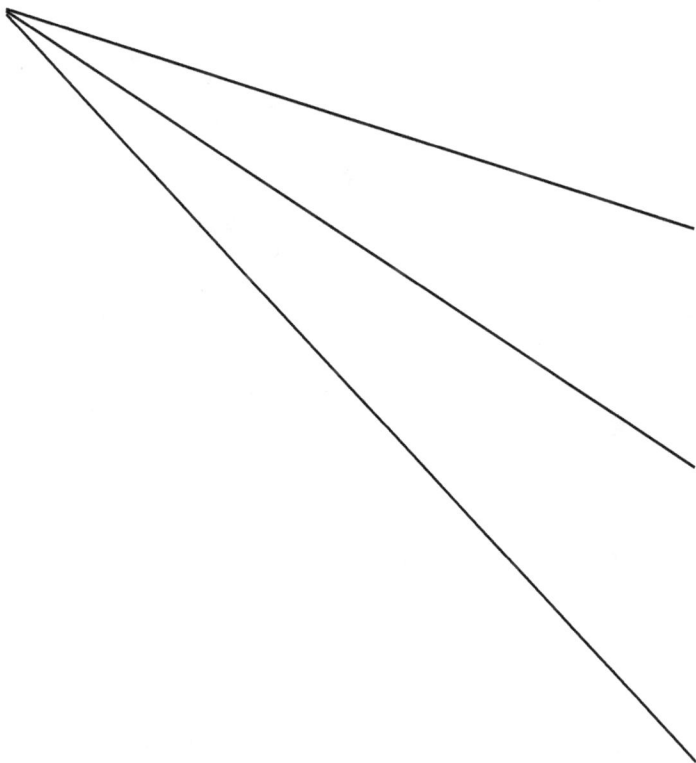

Sycamore
Platanus occidentalis

Seed Plants of the Chessie Trail
John S. and Anne Knox
Revised 2009

We present here a list of the most important seed-bearing plants that grow along the Chessie Trail. The system of organization used here is based upon scientific names for plants and groups of plants, since these technical categories are universally regulated and understood by those with botanical training. A system based upon common names would be impractical if not impossible, as the same plant may have many common names, and the same common name may be used to refer to different plants.

We have followed the customary botanical practice of recognizing within the seed plants three groups: conifers, monocot seed plants, and dicot seed plants. Conifers are usually easily recognized as cone-bearing plants; most have evergreen and needle-like leaves. The remaining two categories are less familiar. The monocot seed plants form a single first seedling leaf (the cotyledon) which is very inconspicuous; they have parallel leaf venation and flower parts in multiples of three. By contrast the dicot seed plants produce two first seedling leaves, have net veined leaves, and have flower parts in multiples of four or five. Within each of these three groups, the conventional botanical practice has been followed of grouping together into families plants sharing reproductive features, such as the numbers and shapes of floral and fruit parts. Alphabetical listing has been used to arrange families within the three groups and genera within families.

One will also find various nonseed-bearing plants along the Trail that are not treated in this list. Plants that reproduce by spores, and which may be found on the Trail, include algae, fungi, mosses, liverworts, ferns, clubmosses, and horsetails. The most conspicuous spore plant along the Trail is the horsetail or scouring rush *(Equisetum hyemale)* which grows abundantly at the very margin of the Trail.

The list of scientific names of plants is accompanied by columns that provide additional information, including the plants' common names, origins, life form, blooming season and habitats.

Origins

The "Origin" column is used to report whether a plant is native ("N") or alien ("A"). Alien plants are native to other parts of the world. Through the inadvertent or intentional agency of man, these plants have been carried here mostly during the past few hundred years. The means of introduction of alien plants has varied. Some were intentionally cultivated as ornamental plants or medicinal herbs. Others produce seed or fruit that readily stick to animal fur or on man's clothing and possessions. No doubt plants of the latter type have hitched rides into Rockbridge County.

Historically, plant seeds introduced inadvertently into ship ballast have been picked up in one country and transferred to another. This is likely one of the ways aliens began their journey to the county. The fact that the Trail follows an old railroad bed and barge canal suggests additional human activities that could have inadvertently disseminated alien plants along the Trail. There exists a particularly interesting pattern of plant distribution along the Trail. The vegetation of the trail roadbed and adjacent trailside is composed of a high percentage (45 percent) of alien plants, most of which are of Old World origin. These alien plants tend to be aggressive weeds; they are well adapted to colonize rapidly and to flourish in chronically disturbed habitats. Continued use and maintenance of this thoroughfare over many years has made it an ideal chronically disturbed habitat for such plants. It is in fact a common pattern for chronically disturbed habitats in the Americas to be heavily populated by alien weeds of Old World origin. Evidence gathered by plant geographers and anthropologists strongly suggests that as man has spread over the globe from the Old World, many weedy plants have evolved and/or expanded their geographic ranges into the disturbed areas that man has created. Given a longer period of association with man over which to evolve adaptations to chronic disturbance, Old World weeds have become some of the world's most aggressive competitors for disturbed habitats.

Life Forms

The "Life Form" column describes the stature and form of the plant.

H = Herb, a plant that is relatively soft and which dies down to the ground each year

S = Shrub, a plant that usually has many woody stems and which grows no taller than five meters

T = Tree, a plant that usually has a single woody stem and which grows taller than five meters

V = Vine, a twining or rambling plant

Flowering Season

The "Flowering Season" column reports the range of months or season during which a plant flowers.

Habitats

The "Habitat" column reports the range of environments in which one may find a plant growing along the Trail. The letter symbols used have the following meanings:

A = Aquatic

F = Floodplain

O = Open fields

R = Rock outcrops and steep slopes subject to frequent disturbances, with a thin and unstable soil

TR = Trail roadbed and immediately adjacent trailside, frequently disturbed by use and maintenance of the trail

W = Upland woods above the influence of river flooding

Comments and References

The "Comments" column reports a varied assortment of additional information such as edibility, economic uses, and distinguishing characteristics.

This list is not designed primarily as a tool for plant identification, although it may aid one in selecting from among the many plants described in plant manuals. We have suggested below several plant manuals that may be used to identify the plants which grow along the Chessie Trail.

Whichever manual is chosen, useful observations in identifying plants include examination of the flower and/ or fruit for structure and number of parts. (*Editors' note*: For example, flowers of the mustard family, including dame's rocket, have four petals, while phlox has five. Seed pods of dame's rocket are long and thin, resembling other mustards in that respect.) The form of leaves and stem should also be noted, along with such additional features as stature and the habitat in which the plant is growing.

1. Elias, T., 2000. *The Complete Trees of North America.* Times. A very useful book for the highly motivated amateur and the professional botanist.

2. Gleason, H. A. and A. Cronquist, 1991. *Manual of Vascular Plants of the Northeastern U.S. and Adjacent Canada,* 2nd ed. New York Botanical Garden. This manual is for the advanced student and professional botanist.

3. Holmgren, N. H., P.K. Holmgren, R. A. Jess, K.M. McCauley, and L. Vogel, 1998. *Illustrated Companion to Gleason and Conquist's Manual.* New York Botanical Garden. This manual is for the advanced student and professional botanist.

4. Newcomb, L. 1989. *Newcomb's Wildflower Guide*. Little Brown. An easy to use manual for the amateur, it does not treat all plants found along the Trail.

5. Radford, A. E., H. E. Ahles, C. R. Bell. 1968. *Manual of the Vascular Flora of the Carolinas*. U.N.C. A technical manual for the advanced student and professional.

Soapwort; Bouncing-Bet
(Saponaria officinalis)

Virginia Day-flower
(Commelina communis)

Update from the field, 2009

As the Knoxes' article states, a substantial proportion of plant species found along the Chessie Trail roadbed and adjacent trailside is made up of non-native plants. Without detailed records over the past century, it is difficult to ascertain if the natives here have existed among the non-natives in proportions similar to what is present now, or if natives are losing ground today.

Lists kept since 2002 by the local chapter of the Virginia Native Plant Society, of wildflowers blooming along the western and central sections of the Chessie Trail, show that there are segments of the trail that exhibit a remarkable diversity of herbaceous plant species. Some populations in deeper shade, or among shady rock outcroppings, are apparently thriving, despite a recent four-year drought and the presence of numerous aggressive non-native species nearby. Other populations appear currently at risk of being overwhelmed by a handful of particularly aggressive non-native species.

In one location, for example, a bank of *Senecio aurens*, golden ragwort, itself a sturdy seed producer and spreader, with an over-wintering rosette, is being progressively curtained by Japanese honeysuckle vining down from the slope above and rooting wherever it touches soil.

A substantial drift of *Trillium sessile,* toad shade trillium, which is growing in the dappled shade of a six-foot spicebush and several years ago had a virtual monopoly on the spot, is now in competition with honeysuckle, garlic mustard *(Alliaria officinalis)* and false strawberry *(Duchesnea indica)*.

West of the South River, there is a half-mile segment of the Trail traversing a pasture between two metal gates. On either side of the Trail, and even here and there on the riverbank among the usual sycamores, are numerous large *Ailanthus altissimus,* also known as paradise tree or tree-of-heaven. In this stretch, *Ailanthus* is more abundant than any other tree, even the native moisture-loving boxelder, *Acer negundo*. Here *Ailanthus* displays its dominating habits – its tendency to sprout multiple stems (which become trunks) from its roots, grow quickly into the canopy, and develop prodigious clusters of seeds, each one within a flat wing (samara) that can ride the wind miles away. Just beyond the western pasture gate, the existing diversity of established moist woodland species and the shade they create, are currently containing *Ailanthus'* aggressive habit, but wherever there is a break in the canopy, *Ailanthus* seeds or root sprouts are present to begin growing into the space.

There are non-natives whose presence along the trail many of us enjoy, such as fragrant *Hesperis matronalis* (dame's rocket) and tasty *Rubus phoenicolasius* (wineberry), but every site occupied by a non-native is another place not open for a native species, and as the non-natives grow in number of species, and number of individuals, more and more of the ever-shrinking habitat available is unavailable for the many native species already in competition. Many non-natives came along with European settlers, and new ones have been moving in ever since, some of which are unfortunately efficient colonizers.

A relatively new threat to the diversity of native herbaceous species along the trail is *Microstegia vimineum,* Japanese stilt grass. Though an annual, *Microstegia* seeds prolifically and comes up in thick stands that can choke out the seeding stage of shorter, spring-blooming species such as *Phacelia purshii* (Miami mist), *Anemonella thalictroides* (rue anemone), *Polemonium reptans* (Jacob's ladder), *Dentaria* (toothworts), *Claytonia* (spring beauties), and *Dicentra* (Dutchman's breeches, bleeding heart). The shaded soil under stilt grass can also compromise the germination of later-sprouting gems such as *Cassia nictitans* (sensitive plant), *Anemone virginica* (thimbleweed), *Aquilegia canadensis* (columbine), *Campanula* (bellflower), *Mertensia* (Virginia bluebells), and *Specularia* (Venus' looking glass). In just the last five years, stilt grass has exploded along several stretches of the Trail, its seeds spreading easily on wind, animal coats, and hikers' shoes. It is now also spreading from the trail farther back into the woods.

Current research explores various attributes, such as allelopathic effects – the production by a plant of substances that are toxic to other plants – that contribute to the colonizing success of some of the most aggressive and persistent non-natives, such as *Ailanthus* and *Alliaria officinalis* (garlic mustard). Do substances exuded by their roots kill the seedlings of some other species, or prevent germination, or interfere with mycorrhizal relationships, such as those between certain plants' roots and symbiotic fungi? How long do these allelopathic substances remain active in the soil after the plant that produced them is removed? Is the success of these non-native species truly due to their multiple reproductive strategies and defenses, or to the lack of any controlling insect or disease vectors that are present where they came from? Or are there other, less evident changes to the surrounding environment that are weakening natives' reproduction and adaptability, even before the non-natives come onto the scene?

Local groups are working to understand the issues raised by the changing plant communities on the Trail, and to determine whether to attempt to limit invasives' spread, or eradicate them from particular sites. In researching best practices, the answer usually amounts to this: For any lasting success, efforts to control aggressive non-natives need to happen early, thoroughly, and sometimes repeatedly.

It may turn out that these non-natives are beyond hope of controlling. But it is incumbent on those of us who enjoy the Trail to learn to identify the more troublesome non-natives, to ascertain how we can help in preventing their further spread, and to participate in control and eradication efforts. We'll be watching to see what adaptation and population changes happen over the coming years. ❦

Peggy Dyson-Cobb
Upper James River Chapter
Virginia Native Plant Society

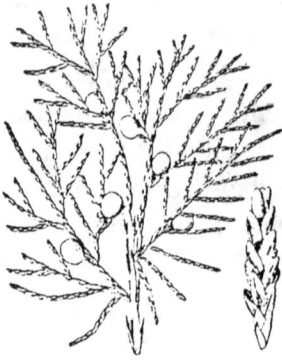

Red Cedar
Juniperus virginiana

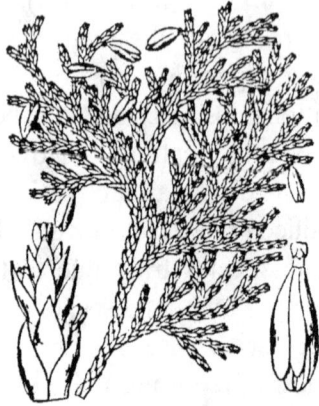

White Cedar; Arbor Vitae
Thuja occidentalis

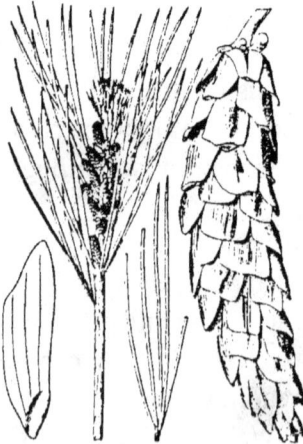

White Pine
Pinus strobus

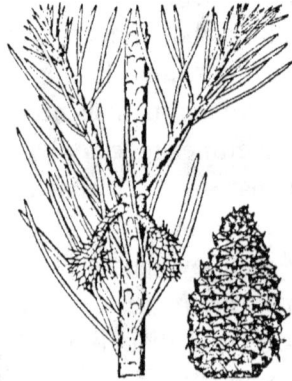

Virginia Pine
Pinus virginiana

Box Elder; Ash-leaved Maple
Acer negundo

Silver Maple
Acer saccharinum

Sugar Maple
Acer saccharum

Black-haw
Viburnum prunifolium

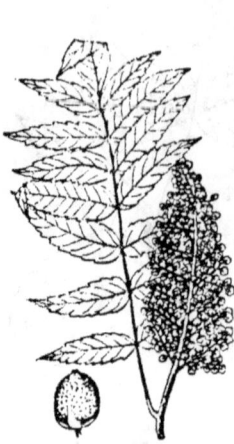

Smooth Sumac
Rhus glabra

Staghorn Sumac
Rhus typhina

Tree-of-Heaven
Ailanthus altissima

Pawpaw
Asimina triloba

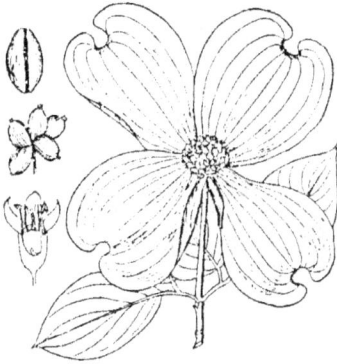

Flowering Dogwood
Cornus florida

Persimmon
Diospyros virginiana

Chinkapin Oak; Yellow Chestnut Oak
Quercus muehlenbergii

Northern Red Oak
Quercus rubra

Bitternut Hickory
Carya cordiformis

Black Walnut
Juglans nigra

White Sassafras
Sassafras albidum

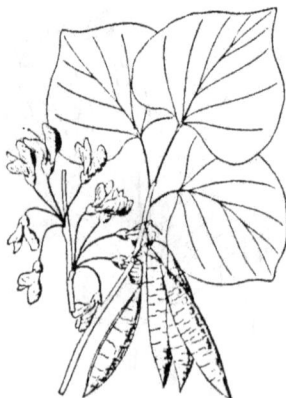

Redbud
Cercis candensis

Black Locust
Robinia pseudo-acacia

Tulip Poplar; Yellow Poplar
Liriodendron tulipifera

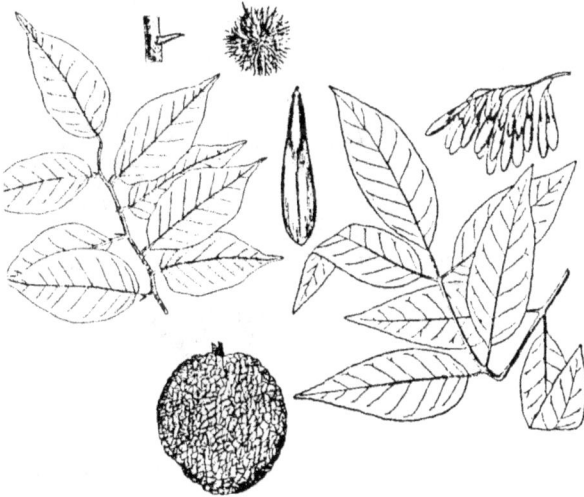

Osage Orange
Maclura pomifera

Ash
Fraxinus americana

Paper Mulberry
Broussonetia papyrifera

White Mulberry
Morus alba

Red Mulberry
Morus ruba

Sycamore
Platanus occidentalis

Black Cherry
Prunus serotina

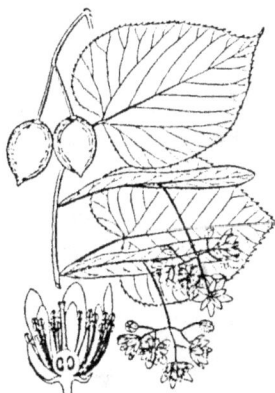

Basswood; American Linden
Tilia americana

Hackberry
Celtis occidentalis

Slippery Elm
Ulmus rubra

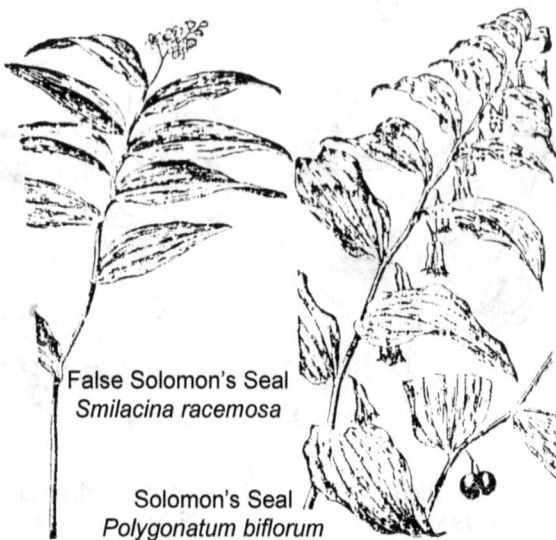

False Solomon's Seal
Smilacina racemosa

Solomon's Seal
Polygonatum biflorum

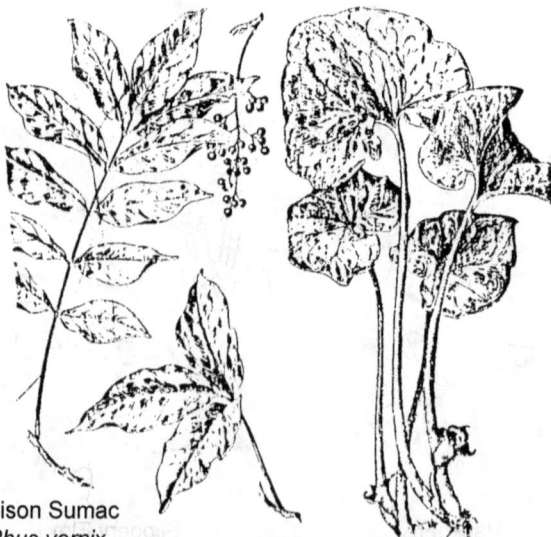

Poison Sumac
Rhus vernix

Poison Ivy
Rhus radicans

Wild Ginger
Asarum canadense

Common Milkweed
Ascelpias syriaca

Butterfly-weed
Asclepias tuberosa

Jewelweed; Pale Touch-me-not
Impatiens pallida

May-apple
Podophyllum peltatum

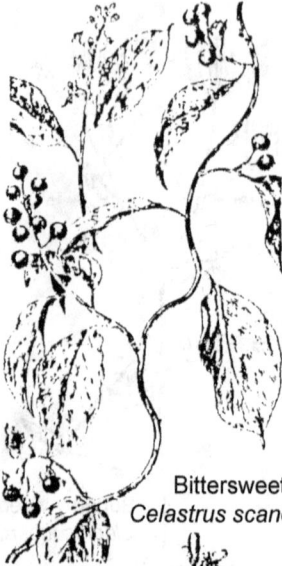

Bittersweet
Celastrus scandens

Common Yarrow
Achillea millefolium

Common Chicory
Cichorium intybus

Canada Goldenrod
Solidago canadensis

Wood Sorrel
Oxalis acetosella

Motherwort *Leonurus cardiaca*

Evening Primrose
Oenothera biennia

Dutchman's Breeches
Dicentra cucullaria

Bloodroot
Sanguinaria canadensis   English Plantain   *Plantago lanceolata*

Moneywort
Lysimachia nummularia

Anemone
Anemone virginiana

42

Leather Flower
Clematis viorna

Violet
Viola papilionacea

Virgin's Bower
Clematis virginiana

Meadow Rue
Thalictrum dioicum

Common Mullein
Verbascum thapsus

Moth Mullein
Verbascum blattaria

Queen Anne's Lace;
Wild Carrot
Daucus carota

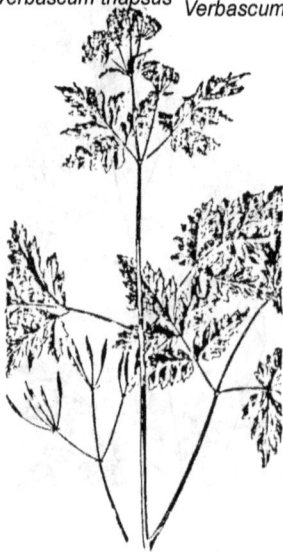

Sweet Cicely
Osmorhiza longistylis

Wild Columbine
Aquilegia canadensis

SEED PLANT CHARTS

Seed Plant charts are found on the following pages. The conifers start at Cupressaceae on page 46. Among the flowering plants, monocots start at Araceae on page 46. Dicots start at Acanthaceae on page 48.

Seed Plants of the Trail

Scientific Name	Common Name	Origin	Life Form	Blooming Season	Habitat	Comments
CUPRESSACEAE	Cypress Family					
Juniperus virginiana	Red Cedar	N	T		R, O, W	Scale-like leaves; blue berry-like cone; very common colonizer of fields
Thuja occidentalis	White Cedar; Arbor Vitae	N	T		R	Scale-like leaves; woody cone; moist slopes; useful in landscaping; used by Indians and settlers to treat scurvy
PINACEAE	Pine Family					
Pinus strobus	White Pine	N	T		W	5 needles per cluster; valuable for lumber and landscaping
Pinus virginiana	Virginia Pine	N	T		W	2 needles per cluster
ARACEAE	Arum Family					
Arisaema triphyllum	Jack-in-the-pulpit	N	H	Mar-Jly	W	Root used by Indians as source of flour after removing calcium oxalate
COMMELINACEAE	Spiderwort Family					
Commelina communis	Day-flower	A	H	May-Oct	TR, F	
GRAMINEAE	Grass Family					
Bromus tectorum	Downy Chess	A	H		O, TR	Very common; visible in early spring in masses; drooping 'flowers'

Scientific name	Common name					Remarks
Eleusine indica	Yard Grass	A	H		O, TR	
Hordeum pusillum	Little Barley	N	H		O	
Hystrix patula	Bottle-brush grass	N	H		W, TR	
Leptochloa fascicularis	Sprangletop	N	H		F	
Microstegium vimineum	Japanese stilt grass	A	H		TR, W	Aggressive annual spreading from trail into woods
Phleum pratense	Timothy	A	H		O	A very important hay grass
Phyllostachys sp.	Bamboo	A	T		F	Dense; stand about 2 ¼ miles from the trail beginning
Setaria viridis	Foxtail grass	A	H		O, TR	
LILIACEAE	*Lily Family*					
Allium cernuum	Nodding wild onion	N	H	Jly-Aug	R	Flower cluster nodding; flowers pink
Allium vineale	Field garlic	A	H	Jun	O, TR	Flower cluster often produces only bulblets; flowers vary from white to red
Asparagus officinalis	Asparagus	A	H	May-Jun	O, TR	The edible asparagus; escaped from cultivation; red berries
Hemerocallis fulva	Common day-lily	A	H	Jun-Jly	F, TR	Flowers orange; escaped from cultivation; buds, flowers, and roots are edible

Scientific Name	Common Name	Origin	Life Form	Blooming Season	Habitat	Comments
Polygonatum biflorum	Solomon's seal	N	H	May-Jun	W	Greenish-white flowers along the stem, mostly in pairs
Smilacina racemosa	False Solomon's seal	N	H	May-Jun	W	Numerous small white flowers at end of stem
Smilax herbacea	Carrion-flower	N	V	May-Jun	F, W	Relative of catbrier and greenbrier but lacking thorns; small greenish flowers in umbels, foul smelling; young shoots used like asparagus
Trillium grandiflorum	Trillium	N	H	Apr-May	W	Petals white, turning pink with age; flower stalked
Trillium viride	Trillium	N	H	Apr-May	W	Petals reddish-purple; flower not stalked; leaves mottled
ACANTHEACEAE	*Acanthus Family*					
Justicia americana	Water willow	N	H	Jun-Aug	A	Petals violet to light blue
Ruellia strepens	Ruellia	N	H	May-Jun	F, TR	Petals blue
ACERACEAE	*Maple Family*					

Scientific name	Common name					Notes
Acer negundo	Box elder	N	T	Mar-Apr	F	Young shoots green; 3-5 leaflets per leaf; common in floodplain; good source of maple syrup
Acer saccharinum	Silver maple	N	T	Feb-Apr	F	Underside of leaf white; a source of maple syrup
Acer saccharum	Sugar maple	N	T	Apr-Jun	W	Excellent source of maple syrup; high quality furniture wood
AIZOACEAE	Carpet-weed Family					
Mollugo verticillata	Carpet-weed	A	H	Jun-Sep	TR	May be used as a potherb
AMARANTHACEAE						
Amaranthus hybridus	Pigweed	A	H	Jun-Sep	TR	Young plants used as potherbs; seeds used to make flour
ANACARDIACEAE	*Cashew Family*					
Rhus aromatica	Fragrant sumac	N	S	Feb-May	O, TR, R	Bushy shrub; red fruit; 3 leaflets per leaf
Rhus glabra	Smooth sumac	N	S	May-Jun	O, TR	Sparsely branched shrub; branches and leaf stalks hairless; fruit red; berries leached in cool water to make acid drink
Rhus radicans	Poison ivy	N	S, V	Apr-May	O, TR	Grayish-white fruit; 3 leaflets per leaf; contact with any part may cause rash

Scientific Name	Common Name	Origin	Life Form	Blooming Season	Habitat	Comments
Rhus typhina	Staghorn sumac	N	S,T	May-Jun	O, TR	Sparsely branched; soft hairs on branches and leaf stalks; fruit red; berries leached in cool water to make acid drink
ANNONACEAE	*Custard-apple Family*					
Asimina triloba	Pawpaw	N	S, T	Apr-May	F, W	Fruit may be eaten raw or cooked in pie or cake
APOCYNACEAE	*Dogbane Family*					
Apocynum cannabinum	Indian hemp	N	H	Jun-Sep	O, TR	Milkweed-like plant; sap milky; fruit composed of paired 10-15 cm slender tubes containing seeds with long white fibers; poisonous
Vinca minor	Periwinkle	A	H	Apr-May	W, O, TR	Blue flowered ground cover planted around home sites
ARISTOLOCHIACEAE	*Birthwort Family*					
Asarum canadense	Wild ginger	N	H	Apr-May	W	Root may be used as ginger substitute
ASCLEPIADACEAE	*Milkweed Family*					

Scientific name	Common name					Notes
Ascelpias syriaca	Common milkweed	N	H	Jun-Aug	O, TR	Flowers vary from purple to green; young shoots, leaves, flower buds and pods may be cooked and eaten like asparagus; raw shoots are poisonous
Asclepias tuberosa	Butterfly-weed	N	H	Jun-Aug	O, TR	Flowers orange-red
Gonobolus decipiens	Angle-pod	N	V	May-Jun	O, TR, W	
BALSAMINACEAE	Touch-me-not Family					
Impatiens pallida	Pale touch-me-not, jewelweed	N	H	Jun-Sep	F, W	Juice is reputed to prevent poison ivy rash; grows in moist, shaded habitats
BERBERIDACEAE	Barberry Family					
Jeffersonia diphylla	Twinleaf	N	H	April	TR	Ephemeral white flowers followed by lidded seed capsules
Podophyllum peltatum	May-apple	N	H	May	W	Fruit used to make jelly and marmalade; other parts, including seed, are poisonous
BIGNONIACEAE	Bignonia Family					
Campsis radicans	Trumpet-creeper	N	V	Jly-Aug	TR, R, W	Large trumpet-shaped orange-red flowers; common along trail; fruit 10-15 cm long, cigar shaped, bearing winged seed
BORAGINACEAE	Borage Family					

Scientific Name	Common Name	Origin	Life Form	Blooming Season	Habitat	Comments
Mertensia virginica	Bluebells	N	H	Apr-May	W, F	Showy clusters of blue to pink pendant flowers
CAMPANULACEAE	*Bluebell Family*					
Specularia perfoliata	Venus' looking-glass	N	V	May-Jun	O, TR	Blue flowers
CAPRIFOLIACEAE	*Honeysuckle Family*					
Lonicera sp.	Bush honeysuckle	A	S	May-Jun	O, TR, W	
Lonicera japonica	Japanese honeysuckle	A	V	May-Sep	F, W, TR	Very common invasive vine
Sambucus canadensis	Common elder berry	N	S	Jun-Jly	F, TR	Berries, purple-black; berries used for preserves, pies, fresh beverage and wine
Symphoricarpos orbiculatus	Coralberry	N	S	Jun-Aug	O, TR	1-1.5 m shrub; red to pink berries attached to along length of stem persist through winter
Triosteum perofoliatum	Horse-gentian	N	H	May-Jun	O, TR, W	Berries used as coffee-substitute
Viburnum acerifolium	Mapleleaf viburnum	N	S	May-Jun	W	

Viburnum prunifolium	Black-haw	N	S	Apr-May	W, TR	Leaf shaped like a cherry leaf
CARYOPHYLLACEAE	*Pink Family*					
Dianthus armeria	Deptford pink	A	N	Jun-Sep	TR, F	0.2-0.6 m tall; narrow opposite leaves; small white spotted pink flowers
Lychnis alba	White campion	A	H	Jun-Sep	O, TR	
Saponaria officinalis	Soapwort, Bouncing-Bet	A	H	Jun-Sep	TR, F	Very common; 0.4-0.8m tall; flowers white to pink; produces poisonous soap when crushed in water
Silene virginica	Fire-pink	N	H	Apr-May	TR, W	Crimson, 5-petaled flowers
Stellaria pubera	Star chickweed	N	H	Apr-Jun	TR	
CELASTRACEAE	*Staff-tree Family*					
Celastrus scandens	Bittersweet	N	V	May-Jun	W, TR	Fruit orange, seed red; poisonous
Euonymus americanus	Strawberry bush	N	S	May-Jun	TR	Heart-shaped seed pod
Euonymus atropurpureus	Burning bush	N	S	Jun	W, TR	Fruit red, seed red; poisonous
CHENOPODIACEAE	*Goosefoot Family*					
Chenopodium album	Lamb's quarters	A	H	Jun-Sep	TR	Young shoots and leaves make tasty potherb
COMPOSITAE	*Composite Family*					
Achillea millefolium	Common yarrow	A	H	Jun-Nov	TR	

Scientific Name	Common Name	Origin	Life Form	Blooming Season	Habitat	Comments
Ambrosia artemisiifolia	Ragweed	N	H	Aug-Oct	TR	Releases much pollen, causing hayfever; 1-1.5 m; leaves much divided
Ambrosia trifida	Ragweed	N	H	Jly-Oct	TR	Releases much pollen, causing hayfever; 1-5 m; leaves 3-5 lobed
Bidens frondosa x bipinnata	Beggar's tick	N	H	Jun-Oct	TR, O	
Centaurea maculosa	Spotted knapweed	A	N	Jun-Oct	TR, O	Flowers pink-purple; very common
Cichorium intybus	Common chicory	A	H	Jly-Oct	TR, O	Flowers usually blue, sometimes white; very common; root used as a substitute for or additive to coffee
Erigeron annuus	Daisy fleabane	N	H	Jun-Aug	TR, O	
Eupatorium rugosum	White snakeroot	N	H	Jly-Oct	TR, W	Growing in shade; poisonous
Galinsoga parviflora	Galinsoga	A	N	Jun-Nov	TR	
Hieracium caespitosum	King devil	A	H	May-Aug	TR, O	
Lactuca serriola	Prickly lettuce	A	H	Jun-Aug	O, TR	A close relative of lettuce
Polymnia canadensis	Leaf cup	N	H	Jun-Oct	W, TR	In moist, shaded places; very common rank smelling plant
Rudbeckia hirta	Black-eyed susan	N	H	May-Oct	TR, R, O	
Senecio aureus	Golden ragwort	N	H	Apr-Aug	TR, W	

Solidago spp.	Goldenrods	N		Jly-Oct	W, TR, O	Yellow flowers; several species grow along the trail
Taraxacum officinale	Dandelion	A	H	All year	TR, O	Potherb
Tragopogon dubius	Goats beard	A	H	May-Jly	TR, O	
Verbesina alternifolia	Alternate leafed stick weed	N	H	Aug-Oct	W, F, TR, O	
Verbesina occidentalis	Opposite leafed stick weed	N	H	Aug-Oct	W, F, TR, O	
Xanthium strumarium	Cocklebur	N	H	Jun-Oct	F, TR, O	
CONVOLVULACEAE	*Morning-glory Family*					
Cuscuta polygonorum	Dodder	N	V	Summer	TR, O	Entire plant yellow-orange, parasitic on other plants
Ipomoea pandurata	Wild potato-vine	N	V	Jun-Sep	TR	
CORNACEAE	*Dogwood family*					
Cornus alternifolia	Alternate-leaved dogwood	N	T	May-Jly	TR, W	Flat clusters of small white flowers; fruit blue-black
Cornus ammomum	Silky dogwood	N	S	May-Jun	F	

Scientific Name	Common Name	Origin	Life Form	Blooming Season	Habitat	Comments
Cornus florida	Flowering dogwood	N	T	May-Jun	W	Attractive white or pink flowers; fruit poisonous; hard, dense wood has been used to make tool handles; bark has been used as quinine substitute
CRUCIFERAE	*Mustard Family*					
Alliaria officinalis	Garlic mustard	A	H	May-Jun	TR	Very common in moist, shaded sites; crushed plant smells like garlic; white flowers; aggressive and allelopathic
Arabis canadensis	Sicklepod	N	H	May-Jly		
Barbarea vulgaris	Yellow rocket	A	H	Apr-Jun	TR, O	Potherb which can be eaten on Saint Barbara's day in December; foliage dark-green, shiny; flowers yellow
Brassica campestris	Field mustard	A	H	May-Oct	TR, O	Potherb; close relative of the turnip and other garden mustards; foliage gray-green, flowers yellow
Dentaria laciniata	Cutleaf toothwort	N	H	Apr-May	TR	
Hesperis matronalis	Dame's rocket	A	H	May-Jun	W	Very common; flowers vary from purple to white

Lepidium virginicum	Pepper grass	N	H	May-Jun	TR, O	Shoots and pods may substitute for water-cress
CUCURBITACEAE	*Gourd Family*					
Sicyos angulatus	Bur cucumber	N	V	Jly-Aug	TR, F	In damp sites; fruit avoid, 15mm long, hairy and spiny
DIPSACACEAE	*Teasel Family*					
Dipsacus sylvestris	Teasel	A	H	Jly-Sep	TR, O	Very common; tall (0.5-2.0 m) herb with spiny flower head
EBENACEAE	*Ebony Family*					
Diospyros virginiana	Persimmon	N	T	May-Jun	W, TR	Edible fruit eaten fresh; also used to make pudding, syrup or vinegar
ELAEAGNACEAE	*Oleaster Family*					
Elaeagnus angustifolia	Russian olive	A	S	Jun-Jly	W, O, F, TR	Aggressive plant covered with silvery scales; sometimes thorny
EUPHORBIACEAE	*Spurge Family*					
Acalypha rhomboidea	Three-seeded mercury	N	H	Summer	TR	
Euphorbia corollata	Flowering spurge	N	H	Summer	TR	
Euphorbia dentata	Poinsettia spurge	N	H	Summer	TR	All parts of the plant are poisonous; milky sap may cause dermatitis

Scientific Name	Common Name	Origin	Life Form	Blooming Season	Habitat	Comments
FAGACEAE	*Beech Family*					
Quercus muehlenbergii	Chinkapin oak	N	T	Spring	TR	Grows on limestone outcrops
Quercus rubra	Northern red oak	N	T	Apr-May	W	Valuable for lumber
FUMARIACEAE	*Fumitory Family*					
Corydalis flavula	Corydalis	N	H	Apr-May	TR, W, F	In moist, shaded areas
Dicentra cucullaria	Dutchman's breeches	N	H	Apr-May	W	Contains poisonous alkaloids
GERANIACEAE	*Geranium Family*					
Erodium cicutarium	Heronsbill	A	H	Mar-Sep	TR, O	Flowers pink to purple; in open sites; leaves used as salad and potherb
Geranium carolinianum	Wild geranium	N	H	May-Aug	TR	In open sites; annual plant
Geranium maculatum	Wild geranium	N	H	Apr-Jun	W	Perennial plant with thick root
Geranium molle	Wild geranium	A	H	Summer	TR	In open sites; annual or biennial
HYDROPHYLLACEAE	*Waterleaf Family*					
Phacelia purshii	Miami mist	N	H	Apr-Jun	TR, W	Small plant with blue, fringed flowers
JUGLANDACEAE	*Walnut Family*					
Carya cordiformis	Bitternut hickory	N	T	Spring	W	Valuable wood for tool handles and fuel

		N	T	Spring	W	
Juglans nigra	Black walnut				W	Valuable lumber
LABIATAE	*Mint Family*					
Glecoma hederacea	Ground ivy	A	H	Apr-Jun	TR, W	Trailing plant with blue flowers; very common
Lamium amplexicaule	Dead nettle	A	H	Mar-Nov	TR	Flowers pink; leaves clasping stem, not stalked
Lamium purpureum	Dead nettle	A	H	Apr-Oct	TR	Flowers pink; leaves stalked
Monarda clinopodia	Horsemint	N	H	Jun-Jly	W, TR	
Nepeta cataria	Catnip	A	H	Jly-Oct	TR	Used to make tea
Perilla frutescens	Beefsteak plant	A	H	Aug-Oct	TR	Foliage often purple
Pycnanthemum tenuifolium	Mountain mist	N	H	Summer	TR	
Salvia lyrata	Lyreleaf Sage	N	H	May-Jun	W	
Teucrium canadense	Germander	N	H	Jun-Aug	W	
LAURACEAE	*Laurel Family*					
Lindera benzoin	Spicebush	N	S	Apr-May	F, W	Many small yellow flowers formed along branches before leaves appear; red berries; crushed bark and foliage have strong, spicy odor

Scientific Name	Common Name	Origin	Life Form	Blooming Season	Habitat	Comments
Sassafras albidum	White sassafras	N	T	Apr-May	W, O, TR	All parts of plant have spicy odor when crushed; tea made from root and bark
LEGUMINOSAE	*Pulse Family*					
Cassia nictitans	Wild sensitive plant	N	H	Jly-Sep	TR	Yellow flowers; pinnately compound leaf folds after being touched
Cercis canadensis	Redbud	N	T	Apr	W	Common small tree; many pink flowers are formed along branches before leaves emerge
Coronilla varia	Crown-vetch	A	H	May-Sep	TR, O	Widely planted as ground cover to reduce erosion
Lathyrus latifolius	Everlasting pea	A	V	Jun-Aug	TR	Very common trailing vine with flowers varying from white to pink to purple
Medicago lupulina	Black medick	A	H	May-Jun	TR, O	
Melilotus alba	White sweet clover	A	H	Summer-fall	TR, O	Grown for hay; poisonous when improperly cured
Melilotus officinalis	Yellow sweet clover	A	H	Summer	TR, O	Grown for hay; poisonous when improperly cured

Scientific name	Common name					Notes
Robinia pseudo-acacia	Black locust	N	T	Jun	W,O, TR	Poisonous; wood valuable for fence posts and firewood
LOBELIACEAE	*Lobelia Family*					
Lobelia siphilitica	Lobelia	N	H	Aug-Sep	F, R	In wet sites; showy blue flowers; poisonous
LYTHRACEAE	*Loosestrife Family*					
Cuphea petiolata	Blue waxweed	N	H	Jly-Sep	TR	Small red-purple flowers; plant covered with sticky hairs
Lythrum salicaria	Purple loosestrife	A	H	Jun-Sept	A	Growing populations on sandbars, riverbanks; aggressive
MAGNOLIACEAE	*Magnolia Family*					
Liriodendron tulipifera	Tulip poplar; Yellow poplar	N	T	May	W	Valuable lumber tree
MENISPERMACEAE	*Moonseed Family*					
Menispermum canadense	Moonseed	N	V	Jun-Jly	W, TR	In moist sites
MORACEAE	*Mulberry Family*					
Broussonetia papyrifera	Paper mulberry	A	S-T	Apr-May	TR	Various birds and mammals compete for the fruit; paper or cloth may be made from the bark
Humulus japonicas	Japanese hops	A	V	Jly-Oct	TR	

Scientific Name	Common Name	Origin	Life Form	Blooming Season	Habitat	Comments
Maclura pomifera	Osage orange	N	T	May-Jun	TR	Branches bear large spines; fruit softball size, composed of many sections; wood orange, makes excellent fuel
Morus alba	White mulberry	A	T	Apr-May	TR, W	Underside of leaf not hairy
Morus rubra	Red mulberry	N	T	Apr-Jun	W	Underside of leaf hairy
OLEACEAE	*Olive family*					
Fraxinus sp.	Ash	N	T		W	Valuable timber tree
Ligustrum sinense	Privet	A	S	Jly	TR, W	Many small blackberries; aggressive
ONAGRACEAE	*Evening-primrose Family*					
Gaura sp.	Gaura	N	H	Summer	TR	
Oenothera sp.	Evening-primrose	N	H	Summer	TR	
ORCHIDACEAE	*Orchid Family*					
Orchis spectabilis	Showy orchis	N	H	May-Jun	TR, W	
OROBANCHACEAE	*Broom-rape Family*					

Conopholis americana	Squawroot	N	H	Spr-Jly	W	
OXALIDACEAE						
Oxalis florida	Wood sorrel	N	H	Apr-Sep	TR	
Oxalis stricta	Wood sorrel	N	H	May-Oct	TR	
PAPAVERACEAE *Poppy Family*						
Chelidonium majus	Celandine poppy	A	H	Mar-Aug	TR, W, F	Shaded, damp sites; saffron colored juice; flowers yellow; plant contains poisonous alkaloids
Papaver dubium	Poppy	A	H	May-Jly	TR, O	Flowers light scarlet
Sanguinaria canadensis	Bloodroot	N	H		W	Flowers white; juice red-orange; plant contains poisonous alkaloid
PHRYMACEAE *Lopseed Family*						
Phryma leptostachya	Lopseed	N	H	Jly-Sep	W	
PHYTOLACCACEAE *Pokeweed Family*						
Phytolacca americana	Pokeweed	N	H	Jly-Oct	TR	Young shoots may be eaten as asparagus, mature shoots and root are poisonous
PLANTAGINACEAE *Plantain Family*						
Plantago aristata	Buckhorn, Bracted plantain	N	H	Jun-Nov	TR	

Scientific Name	Common Name	Origin	Life Form	Blooming Season	Habitat	Comments
Plantago lanceolata	English plantain	A	H	May-Oct	TR, O	
Plantago rugelii	Rugel's plantain	N	H	Jly-Oct	TR	
Plantago virginica x rhodosperma	Hybrid plantain	N	H	Apr-Jun	TR	
PLATANACEAE	Plane-tree Family					
Platanus occidentalis	Sycamore	N	T	Apr-Jun	F	The largest deciduous tree in eastern North America; common in floodplain; bark peeling in large plates leaving white and brown patches on trunk. Indians used sap for syrup and sugar
POLEMONIACEAE	Phlox Family					
Phlox divaricata	Wild phlox	N	H	Apr-May	TR	
Polemonium reptans	Jacob's ladder	N	H	May-Jun	TR	
POLYGONACEAE	Buckwheat Family					
Polygonum sp.	Knotweed	A/N	H	Summer	TR	
Rumex sp.	Dock, Sorrel	A/N	H		TR	Young leaves useful as potherbs
Tovara virginianum	Virginia knotweed	N	H	Summer	W, TR	
PORTULACACEAE	Purslane Family					

Claytonia virginica/ caroliniana	Spring beauty	N	H	Apr	TR, W	White or pink flowers; species distinguished by leaf width
PRIMULACEAE	*Primrose Family*					
Anagallis arvensis	Scarlet pimpernel	A	H	Jun-Aug	TR	
Lysimachia nummularia	Moneywort	A	H	Jun-Aug	TR	In shaded sites
RANUNCULACEAE	*Crowfoot Family*					
Anemone virginiana	Thimbleweed	N	H	Jun-Aug	W, TR	
Anemonella thalictroides	Rue anemone	N	H	Apr-May	W, TR	
Aquilegia canadensis	Wild columbine	N	H	Apr-Jun	R	
Cimicifuga racemosa	Black cohosh	N	H	Jun-Aug	W	
Clematis viorna	Leather flower	N	V	May-Jly	R, TR	
Clematis virginiana	Virgin's bower	N	V	Jly-Aug	TR	Leaves opposite, trifoliate; flowers white, in dense clusters; handling may cause dermatitis
Delphinium ajacis	Larkspur	A	H	Summer	TR	Contains poisonous alkaloids which have caused death of livestock
Hepatica acutiloba	Liverwort	N	H	Mar-Apr	TR, W	On rock outcrops
Ranunculus sp.	Buttercup	A/N	H	Spring	TR	
Thalictrum dioicum	Meadow rue	N	H	Spring	W	

Scientific Name	Common Name	Origin	Life Form	Blooming Season	Habitat	Comments
ROSACEAE	*Rose Family*					
Duchesnea indica	Indian strawberry	A	H	Apr-Aug	TR	Inedible fruit; spreads by rooting stems
Potenilla sp.	Cinquefoil	N	H	Summer	TR	
Prunus serotina	Black cherry	N	T	Apr	TR, W	Fruit used in jellies; valuable furniture wood
Rosa multiflora	Mutliflora rose	A	S	May-Jun	TR	Very fragrant; aggressive
Rubus odoratus	Flowering raspberry	N	V	Summer	TR	Spineless; leaf sycamore-like, flower rose-purple
Rubus phoenicolasius	Wineberry	A	V	Summer	TR	Stems spiny and covered with sticky long hairs
Rubus sp.	Blackberry	N	V	Summer	TR	Stem spiny; the center stalk of fruit remains on stem when ripe fruit is picked
RUBIACEAE	*Madder Family*					
Cephalanthus occidentalis	Button bush	N	H	Jly-Aug	F, A	
Diodea teres	Buttonweed	N	H	Jun-Oct	TR	
Galium sp.	Bedstraw	N	H	Summer	TR	
RUTACEAE	*Rue Family*					

			S-T			
Ptelea trifoliata	Stinking ash; wafer ash	N		May-Jly	R, F, W	On moist, shaded sites; fruits have been used as substitute hops
SAXIFRAGACEAE	Saxifrage Family					
Heuchera villosa	Alumroot	N	H	Jun-Aug	R, W	Damp steep sites
SCROPHULARIACEAE	Figwort Family					
Chaenorrhinum minus	Dwarf snapdragon	A	H	Jun-Sep	TR, O	
Scrophularia nodosa x marilandica	Hybrid figwort	A+N	H	Jun-Oct	TR, W	A putative hybrid of a native plant, *S. marilandica*, with the alien *S. nodosa*, which was introduced to this country in ballast
Verbascum blattaria	Moth mullein	A	H	Jun-Sep	TR, O	
Verbascum thapsus	Common mullein	A	H	Jun-Sep	TR, O	Very common, conspicuous rosette of large, densely hairy, gray-green leaves
Veronica sp.	Speedwell	A/N	H	Spr-summer	TR	
SIMAROUBACEAE	Quassia Family					
Ailanthus altissima	Tree-of-heaven, Paradise	A	T	Jun-Jly	TR	Very common tree of disturbed sites; aggressive and allelopathic
SOLANACEAE	Nightshade Family					
Solanum carolinense	Horse nettle	N	H	May-Oct	TR, O	Spiny stems and leaves; fruit yellow; flower white to blue

Scientific Name	Common Name	Origin	Life Form	Blooming Season	Habitat	Comments
STAPHYLEACEAE	Bladdernut Family					
Staphylea trifolia	Bladdernut	N	S-T	Apr-Jun	F, W	On rich soil in shaded sites; leaves opposite, each with 3 leaflets; fruit a three-parted bladder
TILIACEAE	Linden Family					
Tilia americana	Basswood	N	T	Summer	F, TR	Valuable timber tree; tea may be prepared from the flowers and a chocolate substitute from the seeds
ULMACEAE	Elm Family					
Celtis occidentalis	Hackberry	N	T	Summer	F, TR	Mature bark with very thick corky protruding ridges
Ulmus rubra	Slippery elm	N	T	Mar-May		Inner bark mucilaginous
UMBELLIFERAE	Parsley Family					
Conium maculatum	Poison hemlock	A	H	Jun-Aug	TR	Poisonous; purple spotted stems to 2 m.
Cryptotaenia canadensis	Honewort	N	H	Jun-Sep	W	
Daucus carota	Wild Carrot, Queen Anne's lace	A	H	May-Oct	TR, O	The progenitor of the cultivated carrot
Osmorhiza longistylis	Sweet cicely	N	H	May-Jun	W, F	The crushed plant smells like anise

Sanicula trifoliata	Black snakeroot	N	H	May-Jun		Spherical seedheads with hooked bristles
Zizea aurea	Golden alexanders	N	H	May-Jun		
URTICAEAE	*Nettle Family*					
Laportea canadensis	Wood nettle	N	H	Jly-Sep	W, F	stem and leaves have stinging hairs
Pilea pumila	Clearweed	N	H	Jly-Oct	W, F	Moist, shaded sites
VIOLACEAE	*Violet Family*					
Viola papilionacea	Violet	N	H	Mar-Jun	TR, F	In damp sites
VITACEAE	*Grape Family*					
Parthenocissus quinquefolia	Virginia creeper	N	V	Jun	TR, W, R	
Vitis sp.	Grape	N	V	Summer	W, F	

Cantharellus cibarius

Mycology Along the Chessie Trail
Katie Letcher Lyle
Revised 2009

The Maury River Basin produces mushrooms in great variety; at least two hundred kinds are indigenous to the area. All of the major mushroom families have their representatives here, so an amateur is limited only by the extent of his interest. In dry years one may find only a few, but rainy weather turns the forests and fields along the trail into a mycologist's paradise. Mushrooming is a fascinating pastime, but the consumption of wild mushrooms must be approached with the utmost caution.

To most people, the word "mushroom" means the family of gilled fungi called *Agaricaceae. Agaricus campestris,* the common field mushroom, called "champignon" or "field nut" by the French, is *the*

American mushroom; it and the closely related variety *Agaricus bisporus* are the mushrooms sold fresh and canned in markets and used commercially for soups and sauces. Although many agarics, including the delicious honey mushroom *(Armillaria mellea)* and the popular oyster mushroom *(Pleurotus sapidus),* are avidly collected and eaten, and although many are delicious and safe, this is the group to be shunned by amateurs, as all of the really dangerous mushrooms are gilled, and the edible varieties can be all too easily confused with members of the highly toxic *Amanita* group. This confusion is almost invariably what leads to the approximately one thousand cases of mushroom poisoning reported annually in the United States. It is recommended, therefore, that all gilled mushrooms, no matter how certain the seeker is that he has found a harmless variety, be shunned until the hunter has become skilled at identifying characteristics and habitats, and at taking spore prints. Folk methods of detecting poisonous varieties are, in a word, unreliable.

The genus *Morchella,* by contrast, is a good place for the amateur mushroom hunter to start. Morels are safe for amateurs to hunt and eat because they look like no other fungi, being spongy, vaguely Christmas-tree shaped, and lacking gills. Once identified, they are impossible to mistake. In late April and early May (when oak leaves are the size of mouse ears, according to local lore), several species of *Morchella* (morels) may be found in hardwood forests. Earliest are the *Verpa bohemica* and *Morchella hybrida,* known locally as "snakeheads"; they are followed shortly by *Morchella esculenta,* and occasionally a larger variant, *Morchella crassipes,* can be found. Both the black and white morels are considered choice by the local people, who often will damn all other mushrooms as "toadstools."

The *Polyporaceae,* or mushrooms with very fine, even, spongelike bottoms instead of gills, are also safe for amateurs, as they are readily identifiable and none are unwholesome. (Some, however, are inedibly tough and woody.) They flourish throughout the summer months. The true polypores grow on the sides of old trees *(Polyporus sulphureus,* "the chicken mushroom," and *Polyporus semialbinus)* or on old stumps *(Fistulina hepatica,* "the beefsteak mushroom").

Boletus and *Boletinus* mushrooms are polypores that grow on the ground, with stems, and look from the top like gilled mushrooms, though underneath they have sponge instead of gills.

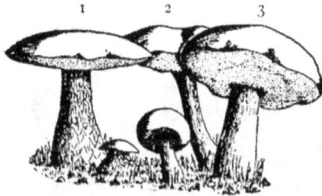

(1) *Boletus edulis (var. clavipes)*
(2 & 3) *Boletus edulis*

Armillaria mellea
"Honey Mushroom"

73

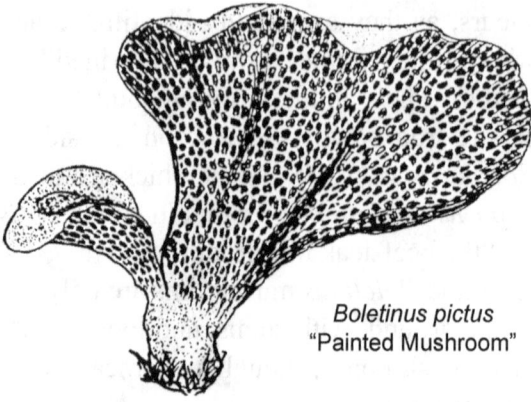

Boletinus pictus
"Painted Mushroom"

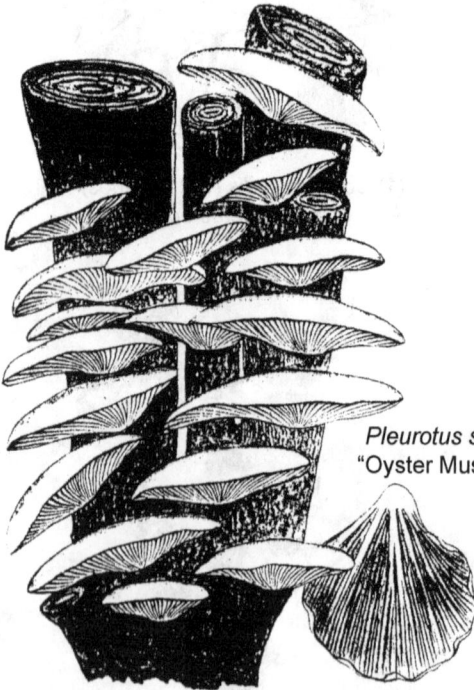

Pleurotus sapidus
"Oyster Mushroom"

More than thirty different *Boletus* varieties have been found in the river basin and along the trail. Some easily identifiable boletes are what the French call "the cepe" and the Italians "porcini," the *Boletus edulis;* "the pinecone mushroom," *Strobilomyces strobilaceus;* "the painted mushroom," *Boletinus pictus;* the *Suillus americanus, Suillus brevipes,* and *Boletus variipes.* Care must be taken with boletes; some authorities label red-bottomed boletes poisonous and some warn against eating boletes that stain blue when wounded, that is, when a piece of their whitish flesh is exposed to air or touch. But with reasonable care and two or three reliable field guides, polypores and boletes are safe for beginners.

The *Clavariaceae,* or coral-like mushrooms, are all safe to eat, though some are bitter or tough, and in general they are not particularly tasty. They flourish in wet summer weather. The choice summer mushrooms, however, are the chanterelles. But these are gilled mushrooms, safe to consume only after a careful study of their characteristically blunt, branching, veiny gills. Chanterelles are placed in the clavaria family by some authorities and the agaric family by others, because of the peculiarity of their gills. *Cantharellus cibarius, Cantharellus cinnabarinus, Cantharellus minoris, Cantharellus craterellus, Cantharellus clavatus,* and *Cantharellus infundibuliformis* are common throughout July, August, and September in rainy years.

In the autumn, the *Calvatiae,* or "puffballs," appear in open fields, at times in great quantities. They are all delicious and safe to eat, but disagreeable unless they are firm, absolutely fresh, and white throughout. They range in size from an inch to about a foot in diameter, the average diameter being one and one-half to four inches. They are a popular local genus; the Maury River Basin has at least eight varieties. Each specimen should be cut in half from top to bottom before cooking to ascertain that it is perfectly fresh and not insect-infested, and also to make absolutely certain that it is not a button form of the deadly *Amanita muscaria,* which fruits at the same time and often in the same location.

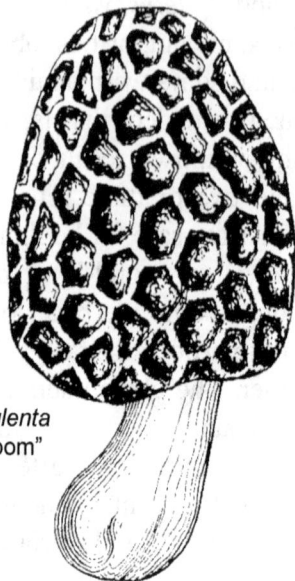

Morchella esculenta
"White Mushroom"

Mushrooms may be used as food fresh or frozen. As they contain up to 95 percent water, drying seems an inefficient method of preservation, and often the mushrooms will mold before drying completely. Since they contain so much water, they freeze well after cooking. A good way to prepare mushrooms for year-round use is to wash them thoroughly, drain them, then "sweat" them in a wide flat pan with a little butter or oil, lecithin spray (like Pam®), dry sherry, or even water. Soon they "collapse" to much less bulk, and the liquid they release can be quickly reduced by evaporation. They can then be frozen in one- or two-cup amounts for future use.

Learning about mushrooms is not difficult. Hunting mushrooms offers a good reason for a pleasant excursion, and eating them is a further reward of the hunt. Mushrooms are low in calories and contain potassium and other minerals, though hardly in significant amounts. But it must be stressed that eating mushrooms from the wild has risks for the uninformed, and for many mushrooms' toxins there are no antidotes known. Therefore, although the morels, the coral mushrooms, the chanterelles, the polypores, the boletes, and the puffballs are in general safer than the large group of agarics, caution must be urged. No one should eat any fungus without considerable study and experience.

Excellent handbooks are Alexander H. Smith, *The Mushroom Hunter's Field Guide* (Ann Arbor: University of Michigan Press, 1980); Louis C. C. Krieger, *The Mushroom Handbook* (New York: Dover, 1967); William Sturgis Thomas, *The Field Book of Common Mushrooms* (New York: Putnam's Sons, 1948); and Orson K. Miller, *Mushrooms of North America* (New York: E. P. Dutton and Co., n.d.).

Update from the field, 2009

Over the years since this guide was first published, I have continued to walk the Chessie Trail in search of mushrooms, and I have continued to find the same kinds of mushrooms as I did 20-some years ago. I have done no scientific "head count" of mushrooms from year to year, but my distinct impression is that, although I find the same kinds year after year, the quantity of mushrooms now is definitely less than it used to be – perhaps because the increased development that Rockbridge County has seen in recent decades has encroached on my hunting areas, but perhaps too because my eyes have grown less sharp with the passing years.

Regarding the books cited at the end of the chapter, they are now out of print, but I continue to use them and recommend them. They are classics, and the persistent hunter will find used copies online and in second-hand bookstores. Mushrooms don't change significantly in the space of a generation, and the information in a good mushroom guide does not need to change either. For those in search of newer imprints, there is a later edition (1980) of Alexander Smith's *Mushroom Hunter's Field Guide*, and Orson Miller has come out with a new book, *North American Mushrooms: A Field Guide to Edible and Inedible Fungi* (Nashville: Falcon Press, 2006). Never having used either of these books, I cannot in good conscience recommend them, but can only express the hope that they measure up to their authors' earlier books. ❦

Katie Letcher Lyle

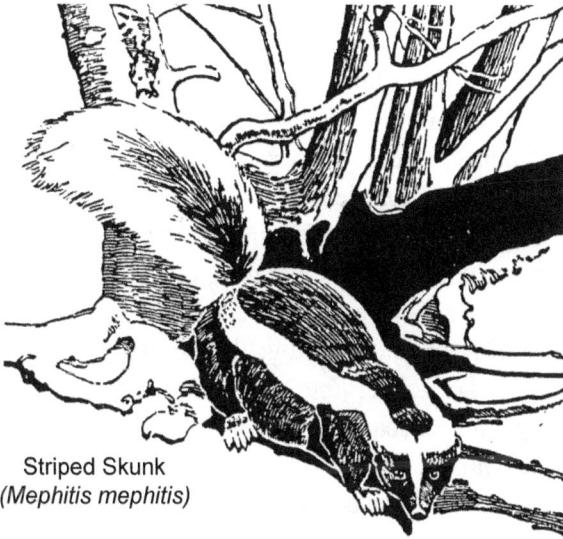

Striped Skunk
(Mephitis mephitis)

Mammals of the Chessie Trail
Michael R. Pelton
2009

Introduction

Streams and their associated riparian buffers are some of the last remnants of good wildlife habitat in a landscape increasingly impacted by human development. Whether mammals are aquatic, semi-aquatic, terrestrial, fossorial, or arboreal, stream corridors often are a logical choice for movement, secure feeding sites, and escape cover. Within the corridor, thick and diverse vegetation as well as woody debris and rock outcroppings enhance habitats for a variety of mammalian prey species; these in turn attract mammalian predators. Also, stream corridors often represent the last vestige of habitat for the safe movements of larger mammals in areas of high human use. These "ribbons of life" therefore tend to concentrate mammals because of the amenities provided by the stream and associated habitats, amenities that are often not found outside the confines of the habitat ribbon.

The Maury River is an excellent example of this ribbon of life and the Chessie Trail offers the opportunity to observe several of Virginia's wild mammals (or their sign). Since many wild mammals exhibit crepuscular (twilight) activity patterns, successful observation is more likely in early morning or late evening; this timing also increases the chances of observing an occasional nocturnal or diurnal species. Because of their shy and secretive nature, it pays to walk slowly and quietly, periodically stopping to listen and watch. In addition, pay particular attention to visible sign of their presence. Scats and tracks are usually the most evident sign and offer the best possibility for species identification. Examine wet muddy spots along the trail for identifiable tracks. Scats may be random but some species prefer to defecate in highly visible locations, such as elevated spots along the trailside or at trail switchbacks and in gaps. Carnivores mark such locations with their scat for identification/territoriality purposes. The size, shape, contents, color, and location of scats are important characteristics to keep in mind when trying to make an identification. Herbivorous mammals leave evidence of their feeding activity such as gnawed nuts, grazed herbaceous vegetation, browsed woody vegetation, flipped rocks, and beaten down berry patches or shrubs.

The Chessie Trail follows a riparian highway used by many species of mammals; this enhances the chances of a hiker finding sign. Hikers also should watch for evidence of wildlife trails. Just like humans, wild mammals are habitual in their use of the landscape in getting from point A to point B. Whether of a mouse or a bear, wildlife trails lace their habitats.

Small mammal trails are best observed in heavy ground cover and along the edge of logs or piles of woody debris. Aquatic and semi-aquatic mammals leave evidence of their access routes to and from the river. Trails to the river may also be those made by some of the larger mammals going to the river to drink. The first step in understanding the behavior and ecology of wild mammals is to practice ways of detecting their presence in the wild.

Wild mammals, whether prey or predator, are naturally adapted to be reclusive and thus, difficult to observe; they have sensory capabilities that far exceed our own, particularly visual, olfactory, and auditory. Consequently, wild mammals are able to detect human presence sooner than we are able to detect them. A combination of habitat selection, nocturnal or crepuscular activity patterns, and coat color – which often acts as a natural camouflage – add to the difficulty of our ability to observe many species of mammals. Their general preference for thick cover is an added handicap for visual observations. One of the best ways to see mammals is to learn to identify their preferred habitats and recognize their sign, such as scats, tracks, marking sites, and vocalizations. A good field guide, particularly for scat and track identification, is a useful item in your field pack. A list of recommended guides is provided at the end of this chapter.

The most recent book on mammals of Virginia lists more than 114 species (Linzey 1998). However, this list includes 26 marine species and nine species (one shrew, one rabbit, four bats, three rodents) not found in the western part of Virginia. Several mammal species in western Virginia are found only at high elevations (one mole, one squirrel, one fisher, three shrews, two rabbits and two or three rodents). Included in the above comprehensive list are cows, horses, and goats as well as species eradicated from our region, such as gray wolf, porcupine, bison, and cougar (Linzey 1998). Subtracting the above species, there are potentially 43 species of wild mammals that may be permanent or occasional residents within the Maury River and Chessie Trail corridor in Rockbridge County, Virginia. Interestingly, feral dogs and cats were not included on the comprehensive list for Virginia. Yet these two species are very common statewide, create significant problems for both humans and other wildlife, and add a dimension of confusion in trying to identify sign (tracks, scats, kills) of similar wild species (fox, bobcat and coyote).

The following mammals are listed phylogenetically, according to their evolutionary relationships. All those listed are not officially documented as present within the Chessie Trail corridor. However, based on range maps, the species' preferred habitats, and the presence of those habitats within the Chessie Trail corridor, the likelihood of their presence ranges from fair to excellent, either as permanent residents or occasional transients.

Marsupials

Virginia opossum (Didelphis virginiana)

The opossum is the only marsupial in North America. This species thrives among a wide array of potential competitors and predators. Facilitating its survival are the opossum's 1) broad food habits, ranging from carrion, garbage, fruits, seeds and eggs to live prey; 2) resistance to some diseases common to placental mammals; 3) "playing possum" in response to a potential threat; and 4) high biotic (reproductive) potential (litters of eight to 12, twice a year). The range of opossums has expanded north into Michigan and New England and the mammal is found in urban, suburban and rural habitats. The common adage "grinning like a possum" is derived from the species' 50 teeth, more than any other North American mammal; when threatened, opossums expose a mouthful of teeth. Opossums are terrestrial and arboreal (assisted by a prehensile tail) and are capable swimmers. Although they are normally nocturnal, daytime activity is not uncommon. This ubiquitous mammal is likely to be seen anywhere along the Chessie Trail.

The Insectivores: Shrews and Moles

Nine species of shrews and three species of moles are listed for Virginia (Linzey 1998). Within the corridor of the Chessie Trail there are likely four shrew species and one or possibly two mole species. As a group, insectivores are very active day and night but are typically buried beneath the leaf litter or under old logs. Members of this group also burrow underground or use burrows made by others. Because of their small size, shrews are preyed upon by a wide variety of reptilian, avian and mammalian predators. On occasion a hiker might find a fresh kill. However, live shrews or moles are seldom seen by hikers. Mammalogists use pit traps to determine the presence of shrews in an area.

Southeastern shrew (Sorex longirostrus)

This shrew is found in a wide variety of habitats ranging from wet riparian forests to dry upland old fields, and is active both day and night. The high metabolic needs of this and other shrews are met by intensive foraging activities. The southeastern shrew feeds on invertebrates such as beetles, crickets, worms and spiders. Life span is short and biotic potential is high, as is true of most small mammals. With intensive trapping, mammalogists would likely find this species along the Chessie Trail corridor in a variety of habitats.

Pygmy shrew (Sorex hoyi)

This shrew is the smallest mammal in North America (2 grams, or approximately .07 oz., about the weight of a nickel) and is widely distributed in Virginia. Pygmy shrews are found in a variety of habitats including cool moist sites near stream courses, deep leaf litter in forested habitats, under and around old logs, and in fence rows and thickets. Food habits are similar to those of other shrews—a wide variety of small invertebrates.

Least shrew (Cryptotis parva)

The least shrew has a shorter tail and is slightly larger than the pygmy shrew. Most shrews are normally intolerant of one another. However, this species exhibits communal behavior; a number of individuals may congregate in nest sites. Owls are a common predator.

Northern short-tailed shrew (Blarina brevicauda)

This is likely the most common species of shrew in Virginia and along the Chessie Trail as well. It is larger than other shrews encountered in the area, and it is the species most likely seen alive or dead along the trail. The saliva of this species is poisonous and is used to capture prey, both vertebrates and invertebrates.

Eastern mole (Scalopus aquaticus)

This insectivore occurs throughout the eastern United States. In Virginia, Eastern moles are ubiquitous in a wide range of habitats. Their search for earthworms and insect larvae just below the soil surface leaves raised furrows, an easily identified sign of their presence. Moles are accused of damaging flower bulbs and roots of shrubs. However, they are almost strictly insectivores. The tunnels they create are used by voles (herbivorous rodents) that do much of the damage to garden or ornamental plants. Because of the strong musky odor, predators often shun Eastern moles as a food source. Hikers can see tunneling activity in all habitats along the Chessie Trail.

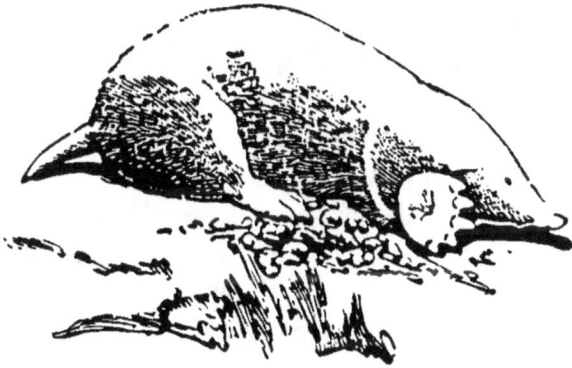

Eastern Mole *(Scalopus aquaticus)*

Star-nosed mole (Condylura christata)

This strange-looking mammal is well-named, as it has 22 fleshy appendages surrounding its nose. Star-nosed moles are not as abundant or as well-known in Virginia as the Eastern mole. However, this species is documented in several locations across the state. Star-nosed moles prefer damp or wet sites along stream courses or in marshy areas. In contrast to Eastern moles, star-nosed moles will sometimes forage in water and above ground. Because of this behavior and the fact that they have no musky odor, they are more vulnerable to predation than Eastern moles.

Bats

There are more than 900 species of bats in the world; North America has 42 species. Virginia is home to 16 bat species; all are insectivores (Linzey, 1998). At least six species occur along the Chessie Trail corridor. Bats are nocturnal and usually observed foraging for insects over openings in the forest and above water. Individual species of bats are hard to identify in flight; a close observation while they are roosting may provide enough information for identification.

While the Trail is typically off-limits to hikers at night, in part to spare nocturnal animals from human contact, bats may be seen in the early evening in those open spaces mentioned above. Their erratic swooping and zigzagging flight is a way of distinguishing them from birds in the twilight.

Handling live bats is not recommended. Technology is now available to identify bats from their echolocation calls during foraging activities. A hand-held unit or "bat detector" receives the ultrasonic frequencies. Each species emits a unique signal or frequency band-width. Because of dramatic declines in bat numbers, several species are protected by state or federal laws. A number of bat conservation practices are recommended, including the construction of bat houses for artificial roost sites. Bats are voracious feeders on mosquitoes and other insects. A single bat can consume hundreds of mosquitoes in a single evening; some species consume from 50 percent to 110 percent of their body weight each night. There are numerous locations along the Trail to observe bats feeding at night.

Species likely to be found in the Chessie Trail corridor:

- Little brown bat (*Myotis lucifugus*)
- Silver-haired bat (*Lasionycteris noctivagans*)
- Eastern pipistrelle (*Pipistrellus subflavus*)
- Big brown bat (*Eptesicus fuscus*)
- Eastern red bat (*Lasiurus borealis*)
- Hoary bat (*Lasiurus cinereus*)

Rabbits and Hares

Five species of lagomorphs ("hare-shaped" species) are documented for Virginia. However, only the Eastern cottontail rabbit occurs in the Chessie Trail corridor. The other four species are located in eastern Virginia (marsh rabbit), the higher elevation mountains (Appalachian cottontail and snowshoe hare) and an introduced population of black-tailed jack rabbits on Cobb Island and adjacent islands off the Virginia coast (Linzey 1998).

Eastern cottontail (Sylvilagus floridanus)

Eastern cottontails are an edge species and prefer habitats where they can find both food and cover in a relatively small area. For preferred cover, cottontails use briar patches, blow-downs, thick fence rows, or slash piles left from harvested trees. Early morning and late evening are the best times to observe feeding activities at the edge of such cover. Both feeding and breeding activities increase in late February and early March. Cottontails are prolific breeders and are capable of producing litters of three to six young every 28 days from March until July. The species is particularly vulnerable to predation by owls, foxes, bobcats, and coyotes. Certainly feral dogs and cats should be added to the list of common predators. Cottontails leave distinct small droppings in piles in contrast to the larger pellets of white-tailed deer.

Rodents

There are more than 2,000 species of rodents in the world. North America has 204 species (Wilson and Ruff 1999). Virginia lists 29 species of rodents (Linzey 1998). The Chessie Trail corridor is likely home to 15 species.

Eastern chipmunk (Tamias striatus)

This small striped rodent is a common woodland species in Virginia and the wooded areas of the Chessie Trail corridor. The five stripes down the back are a distinguishing characteristic. When disturbed, chipmunks make a loud chirping noise, sometimes mistaken for a bird call. Another call made by chipmunks sounds like the cluck of a wild turkey. It is not uncommon to see chipmunks scurrying through the woods with tails held vertically. Chipmunks prefer nuts and seeds but consume a variety of other foods including bird eggs, berries, insects and, unfortunately, bulbs and roots of landscape plantings. This species also collects and caches food for winter and normally hibernates, awakening periodically to feed on food stores or, if necessary, forage outside. Chipmunks are particularly vulnerable to predation by house cats, domestic or feral. Look for chipmunks in woodland areas along the Chessie Trail, particularly in areas with woody debris on the forest floor or abundant understory shrubs and trees.

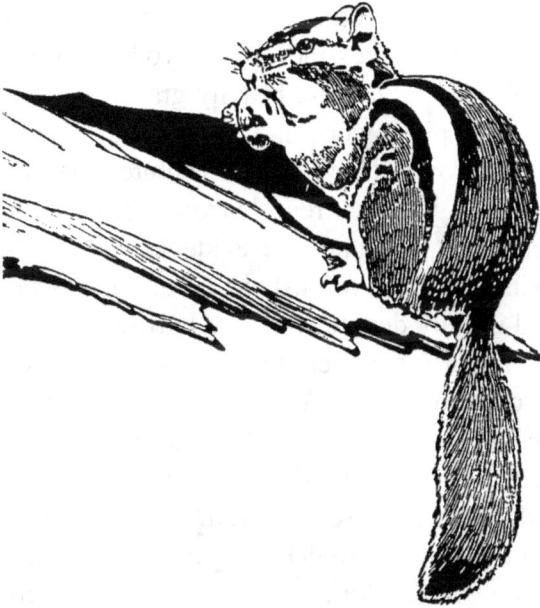

Eastern Chipmunk *(Tamais striatus)*

Woodchuck (Marmota monax)

The eastern marmot, woodchuck, whistle pig, or ground hog is the largest member of the squirrel family in Virginia. Unfortunately, they often are the bane of farmers and gardeners throughout their range. Woodchucks are primarily herbivores and can destroy a small garden almost overnight. Their burrows are often elaborate tunnels and chambers; these burrows sometimes collapse and the hole or depression can damage farm equipment. Foundations of buildings are sometimes undermined and weakened by their excavations. However, the excavated burrows provide dens for a wide variety of other species including skunks, foxes, raccoons, coyotes, and cottontail rabbits. When pastures are heavily grazed, woodchucks move to rock piles, woodland edges, and fence rows to excavate burrows. They seldom venture very far from the safety of their current burrow; trails leading to and from their sanctuary are evident. One- to 2-meter (approximately 3- to 6-foot) circles of dark green grass in a field usually denote an active burrow. After young disperse in early summer, active burrows are occupied by only one woodchuck. Woodchucks can climb trees if threatened. Startled woodchucks also emit a shrill alarm whistle, hence the name whistle pig. Common predators of woodchucks are foxes, bobcat, coyote, and feral or free-ranging dogs. Woodchucks are diurnal; sightings or sign are evident during daylight hours in open areas along the Chessie Trail.

Woodchuck *(Marmota monax)*

Eastern gray squirrel (Sciurus carolinensis)

This tree squirrel occupies the deciduous woodlands of eastern North America, and is the most commonly observed wild mammal in Virginia. Gray squirrels are sometimes called "cat squirrels" because their alarm/warning call sounds like a loud cat meow. For some Virginians, gray squirrels are a prized small game species or a photo opportunity in the backyard. For others they are a nuisance, decimating bird feeders, ruining bird houses, digging up garden plantings, and sometimes damaging wood siding on buildings. Typically, gray squirrels build two types of nests. In summer leaf nests are common. However, in winter tree cavities are preferred. Two litters per year are common, one in late winter and another in summer. Hawks, owls, bobcats and snakes are some of the more noteworthy predators. Gray squirrels are most active in early morning and late evening hours, their favorite feeding times. In fall, gray squirrels are "scatter hoarding"—burying acorns individually in the forest soil. This behavior supplies squirrels with food during winter and serves as a method of seed dispersal. There should be no problem for hikers on the Chessie Trail to see or hear squirrels in wooded areas during their most active feeding periods in early morning or late evening. Leaf nests are common in large deciduous trees.

Eastern Gray Squirrel
(*Sciurus carolinensis*)

Fox squirrel (Sciurus niger)

Fox squirrels are a larger version of gray squirrels. The typical color for fox squirrels in this area is rusty orange with some black/gray around the head. Fox squirrels are less abundant than gray squirrels within the Chessie Trail corridor. Fox squirrels spend more time foraging on the ground and are less nervous and flighty than gray squirrels. Interestingly, the bones of fox squirrels are pinkish yellow due to accumulation of a pigment called porphyrin from their feeding habits. The movement behavior (slow hops) along the ground and the large rusty orange tail give the fox squirrel its fox-like appearance. Fox squirrels tend to prefer more open woodlands and woodland edges than gray squirrels and are active more at midday than grays.

Fox Squirrel *(Sciurus niger)*

Southern flying squirrel (Glaucomys volans)

The southern flying squirrel is nocturnal and does not fly but glides between trees during nightly activities. Because of its nocturnal habits, people live among these squirrels all their lives and never detect their presence. Evidence of their presence is high-pitched squeaks in large trees at night. At dusk they are observed gliding from tree to tree. Glides from tree-tops can exceed 30 meters (98 feet). Unfortunately, flying squirrels take advantage of accessible attics for nesting; their noisy movements and feeding on acorns can be disruptive to people trying to sleep in the room below. Flying squirrels frequently occupy bird houses. The species is communal in winter, with reports of more than 10 congregating inside one house, an obvious energy-saving strategy. The best habitat for flying squirrels is among mature oak trees; such trees provide all their habitat necessities: nesting, resting, feeding, escape, and refuge sites.

Flying Squirrel *(Glaucomys volans)*

Beaver (Castor canadensis)

Once eliminated from much of their historic range in the East, beavers have returned to stream systems in Virginia. Beavers are the largest rodents in North America; a large beaver from Virginia exceeds 20 kg. (44 lb.). Girdling and felling of trees and dam building in large expanses of bottomland forests can have serious impacts on commercially valuable trees. In hilly or mountainous landscapes, damage is more restricted and of less concern. Beaver impoundments have positive ecological benefits ranging from water conservation to enhanced biodiversity by creation of shallow ponds that attract new species to an area. Wherever beavers reside, signs are evident. Along rivers like the Maury, beavers utilize bank dens rather than lodges that are usually constructed in flooded ponds. Besides gnawed and felled trees, dam building activities and beaver tracks in mud at exit/entrance sites into a stream are signs left by beaver activity. Gnawed trees are in evidence along the Trail. Beavers are normally nocturnal but are also active in late evening or early morning.

Beaver *(Castor canadensis)*

Eastern harvest mouse (Reithrodontomys humulis)

This small mouse is nocturnal and occurs in scattered small pockets associated with dense stands of weeds and grasses. This mouse has a tail shorter than its head and body and dark color of dorsal pelage that fades to brown on the sides. The species also does not construct runways; this may lessen the likelihood of detecting their presence in an area. However, the harvest mouse does construct a unique nest. Shredded grass is used as nest material and suspended above the ground in thick clumps of tall vegetation.

White-footed mouse (Peromyscus leucopus)

This is probably the most common mouse in the Chessie Trail corridor. White-footed mice are found in a variety of habitats ranging from agricultural hedgerows to brushy fields to deciduous and coniferous forests and buildings. Nests are a spherical ball made of shredded bark located inside tree cavities, under logs, piles of firewood and lumber. Because of their abundance, they are an important component of the food chain.

Golden mouse (Ochrotomys nuttali)

This unique semi-arboreal mouse has a prehensile-like tail and feet that are smaller than those of terrestrial mice. The above attributes allow this species to spend considerable time foraging above ground in vines, briars, shrubs and small trees. Golden mice are yellowish tan in color (hence the common name) and have white feet. In Virginia, populations of golden mice are scattered in localized pockets and normally associated with riparian habitats with dense understories of honeysuckle, briars, low shrubs or vines. The presence of this rodent is uncertain in the Chessie Trail corridor. However, habitat suitable to its liking is available, albeit in small, scattered locations. Nests are often constructed in thickets two to six meters (approximately six to 10 feet) off the ground.

Allegheny woodrat (Neotoma magister)

This is the only native species of rat found along the Chessie Trail. In contrast to the exotic Norway rat, Allegheny woodrats have large black eyes, silky soft fur, hairy tail and whitish belly. Woodrats are normally associated with rock outcrops, cliffs, boulder piles, wooded bottoms, and old buildings. This species is the eastern version of the western pack rat. Woodrats build middens that consist of whatever items they can carry, including sticks, leaves, pieces of cloth, buttons, glass, pieces of metal, and assorted other trash. Nests are normally quite impressive, consisting of an inner core of shredded bark and leaves and surrounded by a large pile of twigs and leaves; some nests exceed a meter in height and width. Woodrats also use latrine sites rather than defecating in a random or scattered fashion. The nests, middens, and latrine sites are striking evidence of the presence of this rodent.

Meadow vole (Microtus pennsylvanicus)

The meadow vole is the most abundant mammal in the world (Wilson and Ruff 1999). Voles have short tails and small ears. Meadow voles are brown on top and gray beneath. The preferred habitat of this species is thick, moist grassy fields. This abundant herbivore utilizes grasses for both food and cover. Voles make well-defined tunnels through the grass and construct below-ground burrows and nests. Their herbivorous habits and large numbers result in significant damage to valuable shrubs and trees. Damage usually is in the form of girdling the woody base. Just like white-footed mice, meadow voles serve as an important component of the food chain. This species makes up a good proportion of the food habits of a long list of predators.

Woodland vole (Microtus pinetorum)

Woodland voles are a smaller version of meadow voles with even smaller eyes and ears and shorter tail. As the common name indicates, this vole prefers more of a woodland setting with deep leaf litter and scattered thick patches of grass. They also spend considerable time digging underground tunnels and burrows and feeding on plant bulbs, roots, and seeds. Unfortunately, some of these food sources are garden, orchard or ornamental plantings. Because they spend more time below ground than meadow voles, they are not as susceptible to predation and not as likely to be detected by hikers along Chessie Trail.

Muskrat (Ondatra zibethica)

Muskrats are found in salt marshes, swamps, streams, ponds, and lakes throughout Virginia. This herbivore is well adapted to thrive in aquatic environments. Muskrats are large rodents and, when observed swimming, are sometimes mistaken for a beaver. However, the smaller body size and rudder-like, laterally compressed tail is the identifiable key to this species. Because of their thick soft fur, muskrats are trapped for the fur industry. Like beavers, muskrats build lodges in shallow waters of marshes, lakes and ponds. They also dig tunnels and burrows on stream edges; this activity sometimes causes serious damage to pond dams. Muskrats also construct feeding rafts and create swimming channels; both are indications of their presence in an area. Muskrats are typically nocturnal and crepuscular and adapt well to human activities. Raccoons and mink are 2 common predators.

Muskrat *(Ondatra zibethica)*

Norway rat *(Rattus norvegicus)*

This is something of a misnomer, as the Norway rat is an exotic species introduced from Europe, but which is believed to have originated in Japan or Southeast Asia. It is larger than the native Allegheny wood rat, and has a longer, scaly tail. Allegheny wood rats have larger ears and eyes, softer fur, and a haired tail in contrast to the Norway rat. Wherever humans have settled in the world, Norway rats occur. They are extremely adaptable to a wide variety of habitats and are aggressive toward other native rodents. Because of the historic and current settlements of humans along the Maury River, Norway rats are likely present, particularly if garbage and debris are available from close human habitations.

House mouse (Mus musculus)

House mice are exotic introductions originally from Eurasia but are now worldwide in distribution. House mice are dirty gray in color and have a long tapered, bicolor tail. Although the species is typically associated with human dwellings, it also survives in wild populations away from human settlements. Foods of house mice consist of various weed seeds and cultivated grains such as corn and wheat. They also consume animal material such as insects and insect larvae. House mice are regarded as a pest and routinely trapped/poisoned around human settlements. They are susceptible to the usual array of reptilian, avian, and mammalian predators. House mice are found in a variety of natural habitats from fields to woods, as long as there is suitable cover.

Meadow jumping mouse (Zapus hudsonicus)

This unique mouse is readily identified by its long hind legs, large hind feet, and long tail. Meadow jumping mice are found in localized populations. Preferred habitats are near streams and woods edges with associated old fields and brushy cover. They do not make runways but use those made by other species. Their primary food sources are weed seeds, berries, fungi, and insect larvae. Meadow jumping mice are nocturnal and spend several months in hibernation. As their body structure implies, this species moves about in short hops but when alarmed may leap much farther. The presence of this species is questionable for the Chessie Trail corridor; however, some of its preferred habitat is available.

The Carnivores

Carnivores include a number of mammals that also feed on plant sources. In the bear family, for example, species range from the polar bear, which is completely carnivorous, to the giant panda – now generally accepted as a part of the bear family, despite argument to the contrary, and a complete herbivore. Mammals are classified as carnivores on the basis of factors including dentition and their GI tracts.

Red fox (Vulpes vulpes)

Red foxes were likely introduced from Europe into North America in the mid 18th century. As gray wolves and red wolves disappeared from large parts of their former range, red foxes colonized these newly opened habitats. Red foxes are now the most widely distributed carnivore in the world (with the exception of dogs and cats). They are a medium-size carnivore weighing from 3 to 6 kg. (approximately 6 to 12 lb.). Reddish orange body, black feet, and white-tipped tail characterize this mammal. Red foxes prefer a mix of cropland, pastures, brushland, and small woodlots. In recent years, more intensive agriculture and increased numbers of coyotes have resulted in a decline in red fox populations. Primary foods of red foxes are rodents and rabbits. They will dig their own den or modify a woodchuck burrow. This fox and the gray fox adapt to human settlements in suburban areas. The primary predator for red foxes is the coyote.

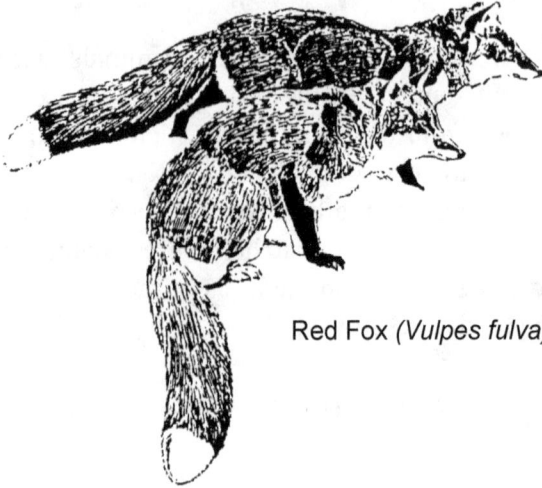

Red Fox *(Vulpes fulva)*

Gray fox (Urocyon cinereoargenteus)

This is the native fox of the eastern United States and Virginia. Gray foxes are about the same size as red foxes; however, they have shorter legs and tail. Color is a grizzled grayish-black body, dark band down the back to the tip of the tail, and rusty orange along the sides. Gray foxes are more reclusive than red foxes and spend more time in woodland habitats. The combined attributes of long, sharp, recurved claws and forelegs that rotate more than other canids allow gray foxes to climb trees to forage for persimmons and other fruit, or to escape potential predators (coyotes). This climbing ability is unique to the gray fox among canids. Den choices include root masses of wind-tilted trees, woodchuck burrows, and brushpiles. Rodents and rabbits are the two most important foods. However, gray foxes are more omnivorous than red foxes and consume such items as grapes, acorns, persimmons, berries, and apples. Their climbing ability and preference for more wooded areas give gray foxes an advantage over red foxes when hunted by coyotes.

Gray Fox
(*Urocyon cinereoargenteus*)

Coyote (Canis latrans)

Coyotes were historically a western species. However, over the past 50 years they have moved eastward via natural dispersal and deliberate capture, transport and release. Coyotes are now distributed throughout Virginia. This adaptable and intelligent canid has exhibited incredible resilience in the face of considerable efforts by state and federal agencies to control or eliminate them. Some regard this "prairie dog" as a useless varmint that kills both wild and domestic animals considered valuable to humans, such as turkeys, deer, grouse, cats, dogs and sheep. Others see coyotes as filling a partial niche left by the loss of wolves and mountain lions from eastern landscapes, and serving a beneficial role in our human-altered ecosystem. Populations of numerous mid-sized carnivores such as skunks, red and gray foxes, opossums, raccoons, and feral dogs and cats are notorious predators on turkey, grouse and other ground nesting animals and far outnumber coyotes in our landscapes. All of the above middle predators are preyed upon by coyotes. So, there are pluses and minuses about the presence of this canid in Virginia. History has shown that we are likely to have this mammal in our rural, suburban, and even urban habitats whether we want it or not. The Chessie Trail corridor has been, is now, and will be used by coyotes as a route during their wide-ranging forays. Because of the extensive use of the corridor by large domestic dogs, accurate identification of coyote sign is sometimes difficult.

Black bear (Ursus americanus)

Black bears are the largest carnivore in Virginia. Since the early 1990s, bear numbers have significantly increased throughout the southern Appalachians. Most of the sightings outside the primary range of the species (state and federal lands and some large private holdings) involve males (usually younger ones) looking for food or a new place to establish a home range. The Chessie Trail corridor is a prime example of a relatively secure way for bears to get from one point to another, particularly under the cover of darkness. There have been a number of bear sightings in and around the Lexington area in recent years. Black bears are intelligent omnivores and are attracted to a wide spectrum of natural and artificial foods. Their primary natural foods are nuts and berries. Unfortunately, they also like seed and suet from bird feeders, dry dog food, greasy outdoor grills, and odors from garbage cans. (Keep in mind that raccoons like these same food items.) The life span of a bear feeding on human food or garbage is measured in days and weeks rather than years, since bears make themselves vulnerable to capture or shooting by frequenting human food sources.

Black bears are typically shy, reclusive wild mammals and want nothing to do with humans. Certainly, an observation of a bear on Chessie Trail would be an unforgettable event; considering recent range expansion of this species into more populated areas of the state, that possibility is enhanced. The footprint of a bear is like no other mammal in Virginia. Large pads and 5 toes are evident if tracking conditions are good. Scats of bears are normally large dark amorphous piles, and berry or nut contents should be visible.

Raccoon (Procyon lotor)

Raccoons occur throughout North America except for northern Canada and Alaska. This species is attracted to edges of streams, ponds, lakes and marshes in both remote and urban landscapes. This mid-sized, primarily nocturnal, and intelligent omnivore usually makes its presence known wherever it resides. While entertaining to some, other people must deal with persistent nuisance activities of the "masked bandit." Adult raccoons weigh 4 to 7 kg. (approximately 9 to 15 lb.). As is true for black bears, raccoons are plantigrade (walk on the flat of their feet); their naked footpads and toes resemble a human print, punctuated with short claw marks. Raccoons are excellent climbers and are attracted to large trees for escape cover, resting places, and winter and natal dens. Winter communal denning is not uncommon. Home ranges are linear when associated with stream systems. Preferred foods include frogs, crayfish, acorns, grapes, eggs of reptiles and birds, human foods, garbage, cat and dog food, and corn. Raccoon scats are smaller versions of bear scats. Coyotes and bobcats are two natural predators.

Least weasel (Mustela nivalis)

This is the smallest carnivore in the world; total length is only 170 to 200 mm. (approximately 7 to 8 inches). Least weasels, like mink and long-tailed weasels, are brown on top and white beneath. Weasels are very reclusive carnivores. Historical records are few for Virginia; many records of this species and the long-tailed weasel are from kills made by house cats. Both species are voracious little carnivores with a body shape that allows them to forage into small tunnels and burrows for mice, voles and shrews. Insects also are eaten. Least weasels sometimes kill more than they can eat (called surplus killing); they cache the extra prey for later consumption. Because of their small size, they are susceptible to predation by larger carnivores such as snakes, owls, and house cats.

Long-tailed weasel (Mustela frenata)

This weasel has a wide distribution throughout the United States, Mexico, Central America and western Canada. Few specimens of the long-tailed weasel are recorded for Virginia. Total length of adults ranges from 300 to 400 mm. This species and the least weasel concentrate their activities around habitats that harbor the most potential prey—brushy areas, areas in woodlands with large amounts of woody debris, and thickets along fence rows. Both the least and long-tailed weasels can be found in the same general habitat type. The long-tailed weasel seldom ventures away from heavy cover, staying under woody debris and in rock crevices in search of small prey. They are known to attack and consume prey more than twice their size, such as rabbits and squirrels.

Mink (Mustela vison)

Mink are a common mustelid in Virginia and lead a semiaquatic existence. They are comfortable on land or in water; time spent in water is associated with hunting prey such as small fish, frogs, crayfish, and muskrats. Mink are residents along most streams in Virginia. This mustelid is substantially larger and more numerous than the two Virginia weasels. Mink are trapped as a valuable furbearer; however, ranch-raised mink have diminished the use of wild mink as a source of fur. Mink is the most likely of the three small mustelids to be observed along the Chessie Trail corridor.

River otter (Lontra canadensis)

River otter were almost totally eliminated from western Virginia because of over-trapping and habitat degradation. Until recently, river otters were classified as a state endangered species. Restoration efforts in various locales in the region have been successful. Otters are the largest of the mustelids in Virginia. They are nocturnal, but sometimes can be seen during the daylight hours. This species is known for its playful activities; they create mud slides along stream banks and repeatedly make trips down the slide and back again. They also stay in family units consisting of the adult pair and young. Dens are typically along the banks of streams under root masses, in beaver or muskrat holes, and occasionally in woodchuck burrows. One unique attribute of this species is the selection and use of latrine sites rather than defecating in scattered locations, a behavior that allows for easy food habits analysis. Preferred foods of river otter are crayfish, slower moving fish, frogs, turtles, muskrats, and a variety of invertebrates. River otters and their signs have been observed on the Maury River in several locations. Given the mobility and extensive home range of this species, they are certain to be using the Maury River along the Chessie Trail.

Striped skunk (Mephitis mephitis)

This medium-sized carnivore is easily identified and well-known, but not particularly well-liked by humans. The distinctive black fur, dorsal white stripes, and bushy tail are sufficient warning that this relatively slow and gentle mammal is a noteworthy adversary. There are two well-developed scent glands on either side of the anus and each contains about a half-ounce of an oily, sulfur-alcohol compound called butylmercaptan. If sufficiently threatened, striped skunks are capable of spraying this powerful scent three to four meters (approximately nine to 12 feet). This successful survival strategy and the broad omnivorous diet of skunks have allowed this species to occupy a variety of habitats, rural, suburban and urban. As long as they are not threatened, skunks maintain an air of "dignified indifference" toward other animals. Adults weigh 0.5 to 2 kg. (approximately 1 to 4 lb.). Dens include woodchuck burrows, under fallen trees, and old stumps. Skunks are normally nocturnal and spend their time foraging for insects, earthworms, small rodents, bird eggs, and berries. Skunks like to forage in open fields; freshly dug small concave holes in the turf are an indication of skunks foraging for insects and grubs. Great horned owls, coyotes, and foxes are their main natural predators. Striped skunks are one species a hiker on Chessie Trail can easily identify by sight and smell!

Eastern spotted skunk (Spilogale putorius)

At the present time it is unknown whether this skunk utilizes the Chessie Trail corridor. This skunk is half the size of the striped skunk. The fur is much finer and denser than the striped skunk. Instead of stripes, white spots are scattered amongst the black and the tail is black with a broad distinct white tip. This species is also nocturnal and more secretive than the striped skunk. It prefers a woodland setting in contrast to the striped skunks' preference for more open habitat. Spotted skunks exhibit a unique behavior of standing on their front feet when threatened; they are capable of spraying in this position but normally spray while on all four feet. Den selection varies considerably and includes woodchuck burrows, tree crevices, and rock piles, or they may dig their own burrow. Spotted skunks occur in scattered localized populations and they are reclusive and rarely observed.

Bobcat (Lynx rufus)

Bobcats are habitat generalists. Because of this trait, bobcats thrive in Virginia and throughout the eastern United States in both rural and suburban areas. The average weight of adult bobcats is 9 to 12 kg. (approximately 18 to 25 lb.). Females are smaller than males. An occasional male will weigh as much as 14 kg. (approximately 30 lb.). Color of the upper body ranges from yellowish-brown to dark brown; the belly is white with black spots. The tip of the short tail (about 160 mm., or 6½ inches) is black on the upper surface and white beneath. There is a white spot in the middle of black fur on the back of the ears. Bobcats occasionally vocalize, particularly during the breeding season; one sound is similar to a human scream. This elusive and mostly nocturnal cat lives amidst human developments without detection by residents. Bobcats prefer rock ledges and crevices for den sites, and take advantage of hollow logs and brush piles for shelter. Their sharp retractile claws allow for easy tree climbing. Food habits are primarily rodents, rabbits, and young deer. Scats and tracks of bobcats are easily identified. Bobcats are protected under an international treaty for the protection of spotted cats. Any harvested bobcat from their entire range in North America must be officially tagged and registered by an appropriate agency; in Virginia that agency is the Virginia Department of Game and Inland Fisheries.

Bobcat *(Lynx rufus)*

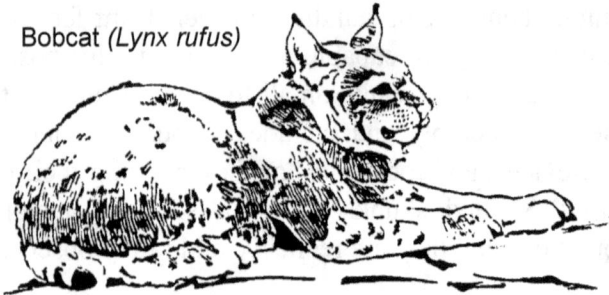

Ungulates
Family Cervidae

White-tailed deer (Odocoileus virginianus)

During the past 50 years, white-tailed deer have returned to the landscapes of Virginia. The recovery of this mammal is a tribute to the organized conservation efforts of state, federal and private groups. It is also an example of the resiliency and adaptability of this popular member of the deer family. Whitetails are now seen in suburban backyards, on golf courses, in gardens and orchards, and all too frequently, dead on roadsides. White-tailed deer have responded dramatically to the diverse patchwork habitats of cropland, brushland, pastures, and woodlots all across Virginia.

Although these deer are browsers on woody and herbaceous vegetation, berry eaters, and acorn feeders in woodlands, they also graze on grassland habitats. Whitetails find secure bedding and fawning sites in thickets, dense grasslands, slash piles left after timber harvests, and patches of brambles. Besides hunting and roadkills, whitetails succumb to various diseases and parasites, and predation from coyotes, free-ranging dogs, and occasionally black bears. Most predatory mortality occurs on fawns in spring. Unless protected, valuable ornamental and garden plants are vulnerable to deer browsing. Hunting is the main mechanism of population control. Deer tracks and scats are easily identified. In the pellet form, individual droppings are larger than the pellets of cottontail rabbits. Since deer have no upper incisors, evidence of browsing on vegetation is more shredded than vegetation eaten by rodents or rabbits that clip vegetation at a 45-degree angle. Whitetails are now a relatively common occurrence in suburban areas surrounding Lexington and Buena Vista and the Chessie Trail corridor.

Whitetail Deer
(*Odocoileus virginianus*)

The Future

Use of the Chessie Trail corridor by mammals is influenced by adjacent habitats; some have a positive influence on the corridor and others have a negative impact. Before further significant changes occur within the corridor and on adjacent lands, baseline surveys should be undertaken to assess the presence, distribution and relative abundance of mammals along the trail corridor. At the present time there are several undocumented species, particularly the smaller and more reclusive ones. A comprehensive live trapping and track station survey would resolve most of the unknowns.

Positive attributes of the corridor for wild mammals include plentiful rocky outcroppings and bluffs and valuable food and cover for mammals—from overstory trees such as red oak, cherry, persimmon, and sycamore, with its sheltering cavities; and from understory growth such as blackberry, pokeweed, mulberry, Russian olive, honeysuckle and grape vines. Grassy floodplains with uncut hayfields and limited inundation provide excellent habitats for small rodents. Also, the corridor is not an island unto itself. Several tributary streams flowing into the corridor serve as connectors to other habitats and travel lanes for mammals.

A primary concern about future viability of mammalian species along the corridor is continued availability of good habitats, particularly for smaller mammals.

Although there are thickets, woody vines, briar patches, and some large woody debris (logs and stumps), these types of habitats are limited in quantity and quality (size, density and composition). Under such circumstances, prey populations will be low and vulnerable. Added to the above issue is the relatively high use of the general area (corridor and adjacent lands) by free-ranging dogs and cats from neighboring developments. Minimizing negative impacts on wild mammals and other animals should be a primary goal of any future developments along the edges of the corridor. ❧

References Cited

Linzey, D. W. 1998. *The Mammals of Virginia.* McDonald and Woodward Publishing, Blacksburg, VA. 459 pp.

Wilson, D. E., and S. Ruff. 1999. *North American Mammals.* Smithsonian Institution Press, Washington, DC. 750 pp.

Other References

Choate, J. R., J. K. Jones, Jr., and C. Jones. 1994. *Handbook of Mammals of the South-Central States.* Louisiana State University Press, Baton Rouge, LA. 304 pp.

Demarais, S., and P. R. Krausman. 2000. *Ecology and Management of Large Mammals in North America.* Prentice-Hall, Inc., Upper Saddle River, NJ. 778 pp.

Feldhamer, G. A., B. C. Thompson, and J. A. Chapman. 2003. *Wild Mammals of North America.* The Johns Hopkins University Press, Baltimore, MD. 1216 pp.

Hall. E. R. 1981. *The Mammals of North America.* John Wiley and Sons, New York, NY. 1181 pp. (2 volumes).

Hamilton, W. J., Jr., and J. O. Whittaker, Jr. 1979. *Mammals of the Eastern United States*. Cornell University Press, Ithaca, NY. 346 pp.

Webster, W. D., J. F. Parnell, and W. C. Biggs, Jr. 1985. *Mammals of the Carolinas, Virginia, and Maryland*. The University of North Carolina Press, Chapel Hill, NC. 255 pp.

Useful Field Guides

Bowers, N., R. Bowers, and K. Kaufman. 2004. *Mammals of North America*. Houghton Mifflin Company, New York, NY. 351 pp.

Burt, W. H., and R. P. Grossenheider. 1952. *A Field Guide to the Mammals*. Houghton Mifflin Company, New York, NY. 289 pp.

Halfpenny, J. C., and J. Bruchac. 2002. *Scats and Tracks of the Southeast*. Globe Pequot Press, Guilford, CT. 149 pp.

Kavanagh, J. 2000. *Animal Tracks: A Pocket Naturalist Guide*. Waterford Press, Inc. Hong Kong, China. Folder.

Kays, R. W., and D. E. Wilson. 2002. *Mammals of North America*. Princeton University Press, Princeton, NJ. 239 pp.

Murie, O. 1954. *A Field Guide to Animal Tracks*. Houghton Mifflin Company, Boston, MA. 374 pp.

Rezendes, P. 1992. *Tracking and the Art of Seeing*. Camden House Publishing, Inc., Charlotte, VT. 320 pp.

Stokes, D., and L. Stokes.1986. *A Guide to Animal Tracking and Behavior*. Little, Brown and Company, Boston, MA. 418 pp.

Black Bear

Coyote

Whitetail Deer

Rabbit

Raccoon

Lynx

Striped Skunk

Wood Duck *(Aix sponsa)*

Birds Along the Chessie Trail
Robert O. Paxton
Revised 2009

About 270 species of birds have been recorded in
Rockbridge County over the years. Many of them might
be found near the Chessie Trail at some season or other,
if one looked long and hard enough. A two-hour stroll
along the Trail in summer or winter is likely to produce
two dozen species or so, and a skilled birder can find
three times that number in spring or fall when migrants
are passing through.

Human activities have changed bird life in the area
now crossed by the Chessie Trail. Before European
settlement, most of the area was timbered, so that
species that preferred open country or brush were limited
to stream-sides, natural clearings, or Indian burns.
Clearing for cultivation in the 18th and 19th centuries

eliminated some forest species such as wild turkey (now thriving again) and the passenger pigeon (now extinct), while it increased the species that live in clearings and edges. Today's mixture of farmland, brush, and woods probably has a greater variety of species than the original homogeneous forest.

Where human use is intensive, however, a few species well-adapted to farmyards, gardens, and city streets proliferate at the expense of natural diversity. These species—mourning doves, rock pigeons (also called rock doves or domestic pigeons), crows, robins, starlings, house sparrows, grackles, and red-winged blackbirds— are what one will see most conspicuously along the Chessie Trail today. Three of these species are not even native: rock pigeons, starlings, and house sparrows were introduced from Europe in the 19th century.

Spring and Fall

Spring and fall are the seasons of greatest variety in bird life along the Chessie Trail. In spring not only do the native summer birds arrive from the south in full song—and bright plumage—but the summer birds of New England and Canada pass through. With hard work and good luck, an expert birder can find sixty species or better in the vicinity of the Trail. The best weather conditions for abundant spring migrants are southwest breezes after the passage of a warm front, particularly if cool rain has bottled up migrants for several previous days.

More birds are present in fall than at any other time. Many young birds have been reared during the preceding summer, and they have not yet suffered the rigors of winter or migration. Many migrants from the north are also passing through. But birds are also hardest to identify in the fall. Young birds may be nondescript, and even adults lose their bright breeding plumage.

The best weather conditions for seeing abundant fall migrants are clearing northwest winds following the passage of a cold front. On such clear, crisp fall days one may see flocks of diurnal migrants like robins, jays, or blackbirds, while nocturnal migrants (flycatchers, vireos, warblers, thrushes) move nervously in the woods, keeping in loose flocks by exchanging calls, feeding to prepare for the long night's flight to come.

Barn Swallow
(*Hirundo rustica*)

Killdeer (*Oxyechus vociferus*)

Summer

During the nesting season, most birds pair off and males sing to proclaim their territories. At this time of year, the attentive hiker can hear even more birds than are seen. Bird activity is greatest early in the season and in the day. At dawn in May or June, hundreds of singing birds form a veritable chorus. At midafternoon in August, however, very little is stirring.

Into early July, begging calls may be heard from young birds being fed by their parents. Particularly patient observers may locate nests with eggs or young.

Remember, though, that a nest to which humans have beaten a path is usually discovered soon by a feral cat or other predator.

About fifty species nest regularly along the Chessie Trail, in addition to the ubiquitous ones mentioned above. No one observer would find them all, of course, in a single day.

Taking the most usual species in the order found in most bird guides, we come first to water birds. A few dark crow-sized green herons live along the river in summer. Canada geese, a non-migratory population descended from captives used as hunters' decoys, have become common in the Maury River Valley. Mallard ducks, some of them semi-domestic fowl from nearby farms, nest near the river, while a few brilliantly colored wood ducks breed in hollow trees along remote stretches.

Both turkey vultures and black vultures lay their eggs in inaccessible crannies in limestone cliffs along the river, while an occasional red-tailed hawk soars over the cliff tops. The American kestrel, or sparrow hawk – a small, brightly colored falcon – is sometimes seen hunting actively over farmland. Like doves and starlings, it is likely to perch on power lines.

Turkey Vulture
(Cathartes aura)

Bob-White Quail
(Colinus virginianus)

120

Kingbird *(Tyrannus tyrannus)*

Sparrow Hawk
(Falco sparverius)

The only shorebird likely in summer is the killdeer, which lays its eggs on the bare ground of open fields. Occasional yellow-billed cuckoos are heard in the forest. Chimney swifts course overhead for insects.

A belted kingfisher or two patrols the river in summer. Downy woodpeckers are the commonest of their family, but in addition to the other normal woodpeckers (flicker, red-bellied, hairy), two or three pairs of the great crow-sized pileated woodpeckers live along the forested limestone cliffs.

Belted Kingfisher *(Ceryle alcyon)*

Among the flycatchers, eastern kingbirds occupy tall treetops over the river, while great crested flycatchers and wood pewees live in dry woods, Acadian flycatchers in damp bottomland forest, and eastern phoebes nest under bridges and eaves. Barn swallows hunt abundantly over farm country.

Carolina chickadees, tufted titmice, and white-breasted nuthatches are common in dry oak woods. House wrens nest around farms, and Carolina wrens in streamside tangles. Mockingbirds sing atop every hedgerow, while catbirds and brown thrashers stay mostly hidden within them. A few wood thrushes nest in the deep woods, and eastern bluebirds are fairly common again where old trees are scattered in the fields.

Carolina Chickadee
(Parus carolinensis)

Carolina Wren
(Thryothorus ludovicianus)

The red-eyed vireo is the commonest vireo, singing in every woodlot, but a few warbling vireos and an occasional yellow-throated vireo are found in tall streamside trees. Black and white warblers and ovenbirds prefer dry woods.

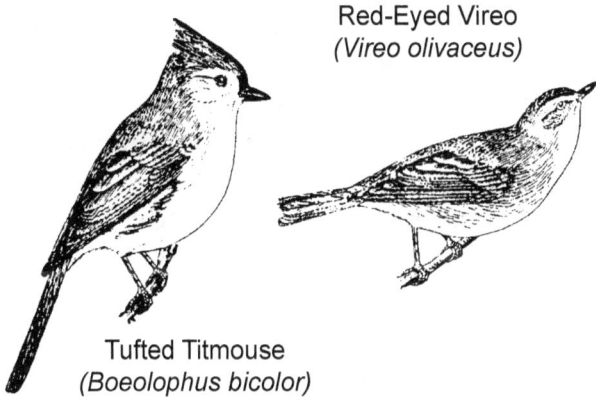

Red-Eyed Vireo
(Vireo olivaceus)

Tufted Titmouse
(Boeolophus bicolor)

Yellow warblers are fairly common in open scattered trees. Large stands of Virginia pine may harbor pine warblers, and overgrown weedy fields, prairie warblers. Every patch of thick streamside vegetation is occupied by a yellow-throat. Louisiana water-thrushes are common along steep stream banks.

Both orchard and Baltimore orioles use tall trees, the latter more frequently at streamside or over roads or lawns. A few scarlet tanagers sing in deep forests.

Cardinals live in every brushy area, and brilliant indigo buntings are the most conspicuous small singer on wires and brushy edges. They must now be carefully distinguished from blue grosbeaks, a larger all-blue bird increasing in this area. American goldfinches nest late, in August, in tall streamside trees. Rufous-sided towhees occupy forested thickets. The commonest resident sparrow is the song sparrow, especially in brush near water, but there are also field sparrows in overgrown pastures and chipping sparrows in isolated shade trees.

Winter

In winter most of our nesting birds have gone south, but there are still many birds around. Though some species depart completely, others unknown here in summer come down from the north. In some cases (robins, for example), the local nesters leave, only to be replaced by others whose behavior on the wintering grounds is quite different. Instead of a familiar pair of robins on each lawn, one may encounter an occasional furtive flock seeking the food and shelter of a cedar thicket.

Like the robins, other wintering birds are no longer singing and many of them are in drabber plumage. They are concentrated in flocks near food, water, or shelter. Areas in between may be quite empty. Forget open country and head for thick cedars, honeysuckle, and weed patches. A birder may walk a long time before hearing a bird call. If one does, a good technique is to make a loud "sh-sh-sh" noise or a high kissing sound on the fingers. If there is a flock of chickadees or sparrows in the brush, curiosity may draw them out into view, especially if the observer is still, or partly concealed.

As long as the river is not frozen, there is a chance to see water birds not found here in summer. A few great blue herons winter regularly along the river. In addition to the summer mallards and Canada geese, there are black ducks and occasional green-winged teal, wigeon, or gadwall. Wary of hunters, they seek out the remotest quiet stretches. There are almost always a few ducks and geese at the mouth of South River. Vultures, red-tailed hawks and kestrels are commoner at this season than in summer. Recently a bald eagle or two has wintered along the river, now that these birds have recovered from the effects of DDT.

Woodpeckers, now joined by yellow-bellied sapsuckers, are common in the forest, but all birds that live on flying or crawling insects, such as flycatchers, vireos, warblers, orioles, and tanagers have gone to the tropics. The exception is eastern phoebes, one or two of which can linger near unfrozen water. Blue jays and crows are much in evidence, and an occasional raven forages away from the mountains.

Chickadees, titmice, and white-breasted nuthatches (reinforced by additional arrivals from the north) form bands and glean for insect eggs or larvae under the bark. They are joined now by red-breasted nuthatches, kinglets and creepers from the north. The only bird likely to sing on sunny winter days is the Carolina wren. Mockingbirds remain common, but catbirds and brown thrashers have departed. Small flocks of bluebirds use scattered trees at the edge of fields, often accompanied by cedar waxwings and myrtle (or yellow-rumped) warblers.

Seed-eating species are generally commoner in winter than in summer. Cardinals now become the most vivid birds of the brush, indigo buntings having gone south. In addition to the familiar goldfinches (now a soberer color), other finches that nest in northern coniferous forests visit this area: purple finches and, occasionally, pine siskins. Purple finches must be distinguished carefully from house finches, a Mexican species that spread into this area in the 1970s from cage birds released accidentally in New York and is now everywhere. Juncos and white-throated sparrows from the mountains now become the most abundant birds of the brush, outnumbering song sparrows. A few white-crowned, field, and tree sparrows also winter in the hedgerows.

Equipment and Reference Works

Anyone who wishes to enjoy the pleasure of finding and identifying birds must expect to spend a lot of time outdoors, to dress correctly for the weather, and to have the patience to study the details of a furtive creature. Binoculars of at least seven power with outer lenses at least 35mm in diameter are a necessity.

It is a good idea to take notes on what you have seen, and on birds you cannot identify. Notes not only help you find the bird later in a guidebook, but also remind you to see all the details.

Anyone serious about identifying birds needs a good field guide. The best one-volume guides are:

(1) *The Peterson Field Guide to Birds of North America* (Boston: Houghton-Mifflin, 2008).

(2) *The Sibley Field Guide to Birds of Eastern North America* (New York: Random House, 2003).

An authoritative annotated list of all species known to have occurred in Rockbridge County up to 1957 is J. J. Murray, *The Birds of Rockbridge County, Virginia* (Virginia Society of Ornithology, Virginia Avifauna No. 1, 1957). This book is out of print, but a copy may be consulted at the Rockbridge Regional Library. Also useful is *Virginia's Birdlife: An Annotated Checklist*, by Stephen C. Rotteborn and Edward S. Brinkley, Fourth Edition (Virginia Society of Ornithology, Virginia Avifauna No. 7, 2007). This book is available from the Treasurer, Virginia Society of Ornithology, 120 Woodbine Drive, Lynchburg, Virginia 24502. ❧

Zebra Swallowtail
(Eurytides marcellus)

Butterflies Along the Chessie Trail
Royster Lyle, Jr.
Updated by Barry Kinzie, 2009

One of the special pleasures of the Chessie Trail is butterfly watching from early spring to late fall. Learning the names and habits of the ones the hiker is liable to see is not all that difficult, and it makes the trip all the more enjoyable.

When pitted against the great variety of birds and the seemingly innumerable flowers and other plants, butterflies are downright easy to identify. There are numerous books that can be useful; two of the most popular are *Butterflies Through Binoculars/ The East* (Jeffrey Glassberg, New York: Oxford University Press, 1999) and *Eastern Butterflies* (Paul A. Poler and Vichai Malikal, Boston: Houghton Mifflin, 1992). On the other hand, beginners may find a smaller book, such as the Golden Guide series, less confusing.

Probably the most common butterfly along the Trail and elsewhere in the country is the cabbage butterfly, also known as the cabbage white *(Pieris rapae)*, which was introduced into North America about 150 years ago in Quebec. Now it is considered a pest by farmers and gardeners because of the great appetite of its caterpillars for cabbages, radishes, and nasturtiums. The cabbage is white with black, dusty wingtips.

In the early spring the orangetip or falcate orangetip *(Anthocharis midea)* is often confused with the cabbage and can be distinguished only by the orange tips on the fore wings. This rather rare butterfly appears for only about two or three weeks in April.

The small yellow butterfly that is almost as common as the cabbage is the common sulphur or clouded sulphur *(Colias philodice)*, which can be seen from very early spring to Thanksgiving and beyond, as long as the weather is fairly mild. The male has a sharp black border around the wings; the female's border is spotted.

The orange sulphur or alfalfa butterfly *(Colias eurytheme)* is the same size as the common sulphur and is easily confused with it. Both the common and the orange sulphurs feed on clover, alfalfa, and vetches, and the two species hybridize and produce many partly orange or white and partly yellow butterflies. Yet the two are still classified as separate species.

A more spectacular sulphur seen along the Trail in mid- to late summer is the large (2 ½ -inch wing span) cloudless sulphur *(Phoebis sennae)*, which breeds farther south but wanders northward through this area. It generally is seen flying very fast and is a brilliant lemon yellow—quite different from the other smaller and slower-paced sulphurs.

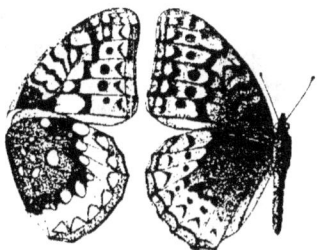

Great Spangled Fritillary
(Speyeria cybele)

Cabbage White
(Pieris rapae)

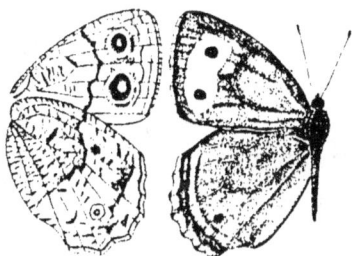

Wood Nymph
(Cercyonis pegala)

Cloudless Giant Sulphur
(Phoebis sennae)

Buckeye
(Junonia ceonia)

Alfalfa/Orange Sulphur
(Colias eurytheme)

The swallowtails *(Papilionidae)* are among the more showy butterflies to be seen on the Trail, and the most common and best known of this group is the Eastern tiger swallowtail *(Papilio glaucus)*. Both males and females are yellow with black tiger stripes and "swallow tails."

Tiger Swallowtail
(Papilio glaucus)

The female tiger swallowtail also has a black form, making it appear very similar to one of the other dark swallowtails that is very distasteful to birds, which assiduously avoid it. This is the survival modification known as Batesian mimicry, not uncommon among butterflies. The key figure in this particular phenomenon is the pipevine swallowtail *(Battus phelenor)*, also known as the blue swallowtail, a dusky-colored butterfly with brighter blue in the area toward the "swallow tail." The distasteful host plants of its caterpillars—such as Dutchman's pipe and Virginia snakeroot—cause the bitter taste in the adult; the taste is so bad and so memorable that birds avoid it. A number of other large dark butterflies gain protection simply because of their color, which falsely warns predators away.

Among the others in the swallowtail group popular along the Trail is the black swallowtail *(Papilio polyxenes)*, the smallest of this series, averaging about 3 inches from wing tip to wing tip. The black swallowtail is attracted to parsley, dill and various carrot plants. The spicebush (or green-clouded) swallowtail *(Pterourus troilus)* is fairly common and takes its nectar from early spring to late summer from joe-pye weed, jewelweed, and honeysuckle, all very plentiful along the Trail.

Probably the most exotic butterfly to be seen on the Chessie Trail is the zebra swallowtail *(Eurytides marcellus)*. Its distinctive long, triangular wings are chalk white with black bands, together with much longer "tails" than the tiger and others. The zebra is often found near pawpaw bushes along the Trail where its caterpillar generally feeds. There are some summers when the zebra is very plentiful. More often it is rather rare.

On any especially warm and sunny day in late March or early April, the hiker may have a chance to see the dark and fast-flying mourning cloak *(Nymphalis antiopa),* which has emerged from hibernation even before the snow has entirely left the ground. It is difficult to appreciate its delicate markings while it is in flight.

Variegated Fritillary
(Euptoieta claudia)

The red-spotted purple *(Limenitis arthemis),* so named for the reddish spots on the underside of its wings, can be found from mid-May on, often alighting in sunny spots in the midst of dark shade.

By midsummer, the rather different common wood nymph *(Cercyonis pegala)* will come bopping across the Trail in its irregular flight a few feet off the ground, usually alighting on a stump or tree trunk. The wood nymph is brown throughout with two blackish dots on a yellow-gray field on the fore wing.

As the summer wears on, three medium-sized butterflies become most prominent, generally alighting with their wings outspread on leaves, flowers, the ground, and sometimes on the hiker's shoulder: the red admiral *(Vanessa atalanta),* the painted lady *(Vanessa cardui),* and the buckeye *(Junonia coenia).* All three have similar habits and can be easily distinguished with a field guide.

Two common members of the fritillary group quite conspicuous during the summer are the great spangled fritillary *(Speyeria cybele)* and the variegated fritillary *(Euptoieta claudia).* Both are orange with black markings on the wings above. The larger of the two, the great spangled, has silver spots on the under side of its hind wings.

Another common butterfly on the Trail is the amazing monarch, or milkweed, butterfly *(Danaus plexippus),* which is found throughout most of North America. This large butterfly is orange with black veins on all four wings and is known for its fondness for flowers. Its flight is slow and sailing, and it can be seen in great numbers in the early fall as it migrates determinedly southward, ultimately reaching Mexico in great numbers. Their remarkable migration was publicized early on by two articles in the *National Geographic Magazine* (April

1963, pp. 588-98, and August 1976, pp. 161-73), which told of how the monarch butterflies make their way by the millions across the Gulf of Mexico to their winter quarters high in the Sierra Madres. These days, Web sites that address their life cycle and migration patterns include a number of National Geographic updates. One site actually describes how they are tagged, as some aspects of their unique journey still remain a mystery.

Viceroy
(Limenitis archippus)

Monarch/Milkweed
(Danaus plexippus)

The caterpillar of the monarch feeds on milkweed (found in abundance along the Trail), which has milky, acrid, poisonous juices, and the adult is therefore highly protected from predators. The "protected" monarch is in turn mimicked in appearance by the less common viceroy *(Limenitis archippus)*, a very similar-looking butterfly from an entirely different family group – again, the Batesian effect.

Two curious little butterflies found along the Trail, members of the angle-wing group, are the question mark *(Polygonia interrogationis)* and the comma *(Polygonia comma)*. The question mark actually has a distinctive, silvery "question mark" (the dot and all) on the underside of the hind wing. The comma in turn has the clearly discernible mark that gives it its name. (The hiker had better learn the more obvious markings and the unusual shape of these two butterflies from a guidebook; the punctuation marks are pretty tough to see, even when the butterflies are at rest.)

There are more than two hundred different skippers in the United States, and a number of them are likely to be present in Rockbridge County. But the one most likely to be seen along the Trail is the silver-spotted skipper *(Epargyreus clarus)*. The silver-spotted is the most common large skipper throughout the country, and like all skippers, its flight is swift and powerful. The brilliant white patch on the hind wing beneath is an easy field mark to spot, even when the butterfly is in flight.

There are several other butterflies that might venture along the Trail, such as the various little blues and coppers and some of the rarer fritillaries. But the above list should be useful to the beginner.

Once considered a polite pastime, butterfly collecting has in recent years been discouraged because of the alarming swings in population figures from year to year in certain species. For instance, in the late 1960s, as a result of the wide use of DDT and other pesticides, many species of butterflies found in Rockbridge County were having a hard time. Things have greatly improved, although there are still inexplicable ups and downs in numbers from year to year. Problems are now arising from loss of habitat due to "clean" farming and development, along with the spread of exotic and invasive plants.

Capturing butterflies today is best done by camera. If you hike the Trail, take pictures and notes, but leave these extraordinary creatures for others to enjoy. Rockbridge County and the Chessie Trail remain a great place to see butterflies, and the new lightweight, close-focus binoculars and cameras make it easy. ❦

Red Admiral
(Vanessa atalanta)

Painted Lady
(Vanessa cardui)

Mourning Cloak
(Nymphalis antiopa)

The Human Impact

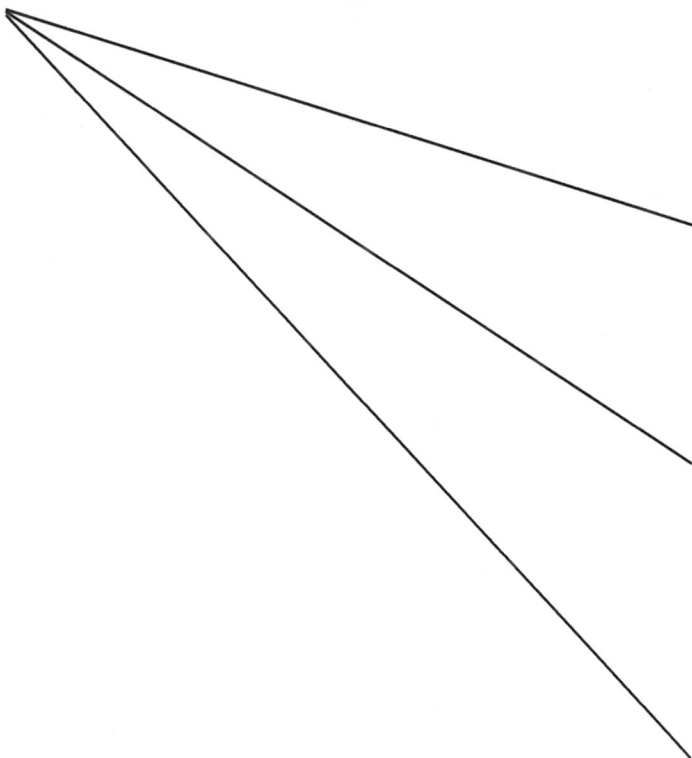

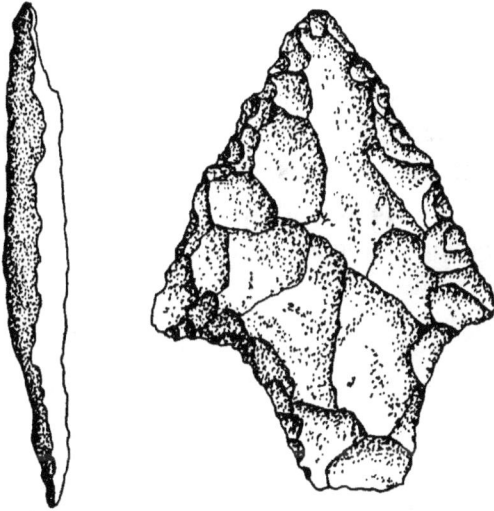

Archaeological Sites Along the Chessie Trail
John M. McDaniel
Updated 2009

Introduction to the 2009 edition

Archaeology is the scientific study of previous civilizations through excavation of sites and the examination of artifacts and cultural remains of past peoples. The Chessie Trail and its environs contain a number of archaeologically significant areas, including prehistoric sites involving Native Americans and the physical remains of historic structures. Some sites are clearly visible to the hiker; the substantial stone walls of a canal lock system still stand after more than a century of inactivity. Other archaeological sites leave no immediately visible evidence, such as prehistoric settlement areas located around the juncture of the Maury and South rivers. Time has faded the evidence of human activity, and, in some instances, significant environmental occurrences like floods have impacted sites to reduce or eliminate them. Whether major or minor, highly visible or long since buried, these areas must be left undisturbed for future generations to study. Pass by these remains with the imagination of the romantic, and with the regard of the reverent.

—*Eliot Balazs, 2009*

In 1978 the State of Virginia, in response to the mandate provided initially by the Historic Preservation Act of 1966, initiated a program to inventory its archaeological sites. Prior to 1978 only a few sites had been recorded in a systematic fashion, and the overwhelming majority of these were from the eastern part of the state. With the establishment of the Washington and Lee Regional Preservation Office as a branch of the Williamsburg-based Virginia Research Center for Archaeology in 1978, the program was implemented in Rockbridge County. Between 1978 and 1988, approximately two hundred prehistoric and historic sites in Rockbridge were recorded. This inventory of sites, and the research that produced the inventory, established a data base from which the history and prehistory of the area continues to be interpreted.

(Ed. note: The Washington and Lee Regional Preservation Office work, which was federally funded at the time, was discontinued after funding cuts in the 1980s. Some of the artifacts and other data collected during the studies described below may be viewed by scholars or school groups by appointment. For more information, scholars or schools should contact the Washington

and Lee Sociology and Anthropology department. For additional information, the Archeological Society of Virginia, Upper James River chapter, may be of assistance. At this printing, the Society's website may be accessed at: *http://asv-archeology.org.*)

A Brief Summary of the Archaeology of Rockbridge County Prehistoric Sites

Prehistoric use of the Rockbridge area in general is characterized by a predominance of Archaic period (roughly from 8000 B.C.E. to 1000 B.C.E.) transient campsites, including some fairly large camps situated on the James River and its larger tributaries. In addition a much lighter frequency of Woodland period (from about 1000 B.C.E. until European contact) occupation exists. Isolated fluted projectile points found within the county provide evidence of early (pre-8000 B.C.E.) Paleo-Indian Native American occupation. The preponderance of Archaic sites in Rockbridge is in stark contrast to other areas of the state, where Woodland period occupation is found with much greater frequency.

Archaic period sites were occupied by transient hunting and gathering groups and are characterized archaeologically by evidence of projectile points and other lithic tools made primarily from chert, quartzite, quartz, siltstones, and greenstone. Sites frequently exhibit firecracked rocks representing hearths or clusters of hearths, a scattering of tools, and waste flakes produced in their manufacture.

Woodland peoples generally practiced agriculture, and as a result became increasingly sedentary. Their sites, especially the larger ones, are characterized by evidence of house patterns identified by post molds, trash and storage pits, and burials. Lithic artifacts are supplemented by ceramics. As mentioned previously, none of these kinds of sites have been identified in the Rockbridge area. It would appear that this portion of the James River watershed most likely formed a buffer zone between competing tribal groups to the east and west.

Historic Sites

From a chronological perspective, European settlement was initiated in the middle of the 18th century. This early settlement was limited to the most accessible zones within what is now the county. These earliest sites, which are far outnumbered by later nineteenth century sites, are limited to the floodplains and natural corridors of travel. As many historians have documented, this first wave of historic immigration comprised the Scots-Irish who came down the Valley from Pennsylvania. Our research has indicated that most of the county was settled between 1820 and 1880, later than some local lore has held. Many sites in the more isolated sections of the county have an initial occupation that at the earliest is between 1820 and 1850.

The very late stages of the 18th century, from about 1780 on, and the entire 19th century saw the establishment of the overwhelming majority of the commercial and industrial sites within the county. Some, such as early mills, have since been abandoned, but many of the original commercial and industrial sites have seen continued use or adaptive development right up to the present.

One of the primary reasons historic archaeology is critical in allowing us to achieve an accurate perception of what transpired in Rockbridge is that most of the extant documentary histories—and the primary documents that provide the basis for those histories—deal with either the earliest stage of settlement or the towns in the area.

Archaeological research has allowed us to gain insights into what the common man was doing in all areas of the county from its earliest settlements on through less documented eras.

From the perspective of the Chessie Trail, while some of the historic sites along the Trail – such as those associated with the canal system – enjoy an extensive history, others such as the domestic structures surveyed can provide us with clues to the culture, lifeways, and adaptive patterns of individuals whose lives are not frequently addressed in detail in the pages of extant histories.

The Chessie Trail and 1982 Work and Findings

The Rockbridge area is geographically broken into three broad physiographic zones: the Blue Ridge Mountains forming the eastern extent of the county, the ridges and valleys of the Allegheny chain to the west, and the area in between, which consists of alluvial valleys and rolling uplands. The major watersheds consist of the James and its tributaries, notably the Maury and South rivers and Buffalo Creek, together with the numerous smaller streams that drain into these tributaries.

From an archaeological perspective, the area adjacent to the Chessie Trail can be broken into two distinct microenvironments: the alluvial floodplain and the rolling uplands. The floodplain width varies, from only a few yards wide to several hundred. Rolling hills and linear cliffs and mature terraces extend outward from the floodplain. Several springs and small streams flow into the Maury along the course of the Trail, but the most elaborate terracing and alluvial deposition occurs around the confluence of the South River and the Maury.

Historically, except for the land used by the James River Canal system and later the Chesapeake and Ohio Railroad, the area has enjoyed intensive agriculture. Today, pasturage and haying are primary land uses. Soils are generally fertile, consisting of silting loams found on the floodplain and clay loams on the older terraces and uplands.

The research strategy for investigating archaeological resources along the Trail during the studies done by the W&L Regional Preservation Office consisted of a systematic survey of the area on the north bank of the Maury, extending into the uplands within sight of the Trail. State site files were consulted that indicated the presence of three prehistoric sites within the immediate area of the confluence of the Maury and South rivers. In addition, the initial reconnaissance indicated the presence of several possible structural remains found in association with Reid's Dam, the South River Lock, and Zimmerman's Lock.

After the initial reconnaissance of the area, the Trail and surrounding environs were divided into 14 segments and assigned to students participating in an archaeology course at Washington and Lee University.

Each segment was then broken into strata or environmental zones that included floodplain, older terrace, cliff, ridge, and uplands. All visible or known features— such as foundations, locks, and prehistoric sites already located—were also designated as strata for testing purposes. Once the strata were defined, a research strategy was employed that included a systematic intensive surface search and subsurface testing of each stratum. Each student was responsible for identifying a datum, or permanent reference point, in order to locate each test area and feature precisely.

Three prehistoric sites near the juncture of the Maury and South rivers had already been inventoried by the W&L Regional Office prior to the student survey. These sites were determined during the survey and research period to represent the most intensive prehistoric use within the immediate area. Presented below is a general discussion of these sites, with the letter and number site designations assigned to them by the Virginia Research Center for Archaeology, for the use of scholars pursuing further research in the area.

1) 44RB34—Primary Terrace Camp

This site is located on private property just west of a residence on the north bank of the South River and encompasses an area roughly 160 feet by 100 feet. Early Archaic through Late Woodland artifacts are represented. A surface collection of the site revealed a small thin triangular point and a Late Archaic projectile point. The landowner has collected from the site in the past and recovered a number of stone tools, including a ground quartzite axe and numerous projectile points.

2) 44RB57—Large Transient Camp

This site off County Road 608 in a southern floodplain extension was approximately 440 by 180 feet. A number of Archaic period artifacts were collected here, including projectile points and tools from the Guilford (Middle Archaic) and Savannah River (Late Archaic) phases. At the time, some flood damage was evident as exhibited in flood-scarred channels within the field. At least two major flood events since that time have likely altered the immediate terrain.

3) 44RB60 —*Large Transient Camp or Base Camp*

This site was located on a well-defined secondary terrace on the south bank of South River. A 1979 alteration of Route 608 had destroyed portions of the site. The site was re-visited through time by Paleo-Indian Native American groups, with points from early, middle, late Archaic and middle Woodland noted.

The survey indicated that the primary prehistoric activity in the area was confined to the Maury-South River confluence but that moderate use was made of a large area in the uplands adjacent to the Trail. Several chert waste flakes and a core fragment were found within shovel test units. The material indicated that a small prehistoric site was probably located nearby, but as no diagnostic artifacts were found, the cultural affiliation could not be determined.

A more extensive flake scatter was found approximately 300 yards away. This site (identified on the survey site map as being in Section 11, site p-2) had not been previously recorded by the W&L Regional Office, and again no diagnostic artifacts were recovered during the survey. This site is in close proximity to 44RB60 and may be indicative of horizontal deposition due to flooding rather than another distinct site.

Another prehistoric site studied was on private property within Section 13 (location p-3). While the survey produced no artifacts, the homeowner indicated that during the construction of his home he collected five projectile points within the immediate vicinity. Unfortunately these points were not available for study. The significant attraction to this area over the centuries would have been a large, strong spring located just off the Trail on the floodplain, near the site location. No doubt other evidence of the site existed, but was likely eradicated when the landowner bulldozed the entire floodplain sometime around the spring after the 1969 Hurricane Camille flood.

The research done in the course of the student survey confirmed that the Maury River and immediate environs within the surrounding uplands attracted prehistoric peoples to the area. Again, the core area of the evidence for activity was concentrated around the confluence of the Maury and South rivers where heavily alluvial deposits formed an elaborate series of terraces. This also holds true for other larger drainages in Rockbridge, where sites are usually found in association with smaller tributaries or large spring sites in close proximity to the rivers. Moderate use was made of the uplands, and in all probability the existing sites were transient camps where tools were manufactured and which were the basis for short-term hunting activities.

The Maury-South River confluence core area is worthy of more intensive investigation and could provide insights concerning the lifestyle and cultural adaptations of prehistoric peoples in Rockbridge County. The presence of an intact midden within the plowing zone of Site 44RB34 was particularly interesting. This site could be worthy of a National Register nomination, as intact middens occur infrequently within Rockbridge. Further testing and research of the site could provide an understanding of man's use of the Maury and its environs over a span of thousands of years.

Historic Evidence

The industrial-transportation sites along the Chessie Trail have already been documented and exhibit visible remains that can be observed and studied by hikers without damage to the archaeological remains of these sites. They are Reid's Lock and Dam, South River Lock, and Zimmerman's Lock.

Two distinct foundations have been found in association with Reid's Lock and the South River Lock, which may be either lock keepers' houses or maintenance buildings associated with the locks. Neither structure produced enough artifacts to enable us to make an interpretation as to precise function or dates of occupation. It is safe to assume, from the foundations' close association and structural similarities to the lock and dam complexes, that they were constructed during the mid-19th century.

The foundation at Reid's Dam consists of a shallow depression lined with cut limestone, which is presumably a cellar. Very few artifacts—primarily scrap metal, window glass, and nails—were recovered. This structure is near the lock, within twenty-five feet of the Trail.

The structural remains near the South River Lock are more elaborate, consisting of a springhouse and a substantial cut-limestone foundation located within a few feet of the Trail. The stone used for the foundation is very similar to the size and shape to the limestone used to construct the lock and dam. The foundation is 25 feet wide and 50 feet long. These dimensions correspond favorably to a photograph of a lock keeper's house found during our research. They were generally rectangular structures. Again very few artifacts were found in association with the foundations, probably because of repeated severe flooding of the Maury.

No structural remains were discovered in association with Zimmerman's Lock. The landowner indicated that after the 1969 flood the entire floodplain around the lock was bulldozed. This disturbance probably destroyed any existing remains.

Several other structures were located during the survey, but overall very few artifacts were produced. The study concluded that severe flooding that has occurred periodically has been catastrophic to the archaeological resources found within the floodplain. The one exception was in the higher areas around the confluence of the South and Maury rivers. Although some flood scarring was visible within that area, the larger floodplain and terraces had escaped the degree of destruction that the rest of the Trail had experienced. The high cliffs and very narrow floodplain found throughout most of the area have repeatedly channeled floodwaters over time so that all sites along the Maury upstream of the juncture had been destroyed before the time of the survey.

Landowners have played a part in the disturbance to sites along the Trail, a case in point being the section of the floodplain adjacent to Zimmerman's Lock that was bulldozed following the destruction wrought by Camille in 1969, when fill was brought in to replace the washed-away topsoil.

Other historic sites identified during the survey include:

A cut-limestone foundation roughly 75 by 25 feet was found within Section 2, 400 feet north of the Trail in the vicinity of the original 1-mile marker. Test pits produced a number of cut nails, several pieces of stoneware, and some glass. The nails are pre-machine cut and date to the early 19th century. This structure may have been a barn or outbuilding associated with one of the upland farms.

Directly opposite Reid's Dam, the remains of what appeared to be a homestead that burned were discovered. Test pits and a surface search revealed charred timbers, the remains of a concrete foundation, and tin roofing material. This site is of 20th century origin.

About a quarter mile past the old 3-mile marker, a rectangular cut-stone foundation measuring 51 by 30 feet was found on the river side of the Trail, with foundation walls one and one-half feet wide and mortared. Three test pits dug within the structure produced no artifacts. A thin layer of humus overlying orange clay was noted within the structure. It may have been associated with river navigation or the railroad, but severe flooding has made more comprehensive interpretations impossible.

A landowner in the area south of the South River bridge informed us that a house stood adjacent to the railroad grade. He also indicated that the house burned and the flood in 1969 decimated the site. No historic period artifacts were discovered.

W&L's designated Section 13—between old Trail markers 6 and 7 near the Buena Vista end of the Trail —held some promise for prehistoric sites, and one was located on land belonging to the Hill family. The landowner also indicated that a house had formerly stood near the Trail, but he could not pinpoint its exact location. The general area he indicated was tested and produced only one piece of bottle glass and a stoneware fragment. No structural remains were extant.

The archaeological survey of the historic use and occupation of the Chessie Trail environs was disappointing in that no significant historic sites were found during the survey. The sites identified were in all probability farm-related, with the exception of the lock-canal complex. The surveyors, however, did document several prehistoric sites, as well as gaining a better understanding of the impact of the geologic folding and of recurrent disturbances within a riverine environment.

For the historic era along the Maury and its tributaries, the documentary record may provide the most concrete information, yielding clues to the overall makeup and cultural activities associated with the operation of the canal and railroad. ❧

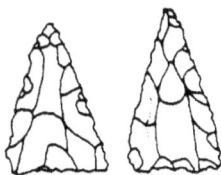

Woodland Triangle

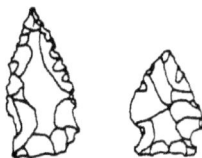

Early-Middle Woodland Notched and Stemmed

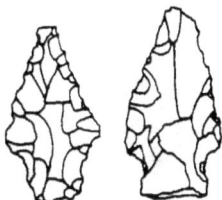

Early-Middle Woodland Stemmed

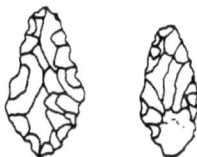

Piscataway-Rossville (Woodland)

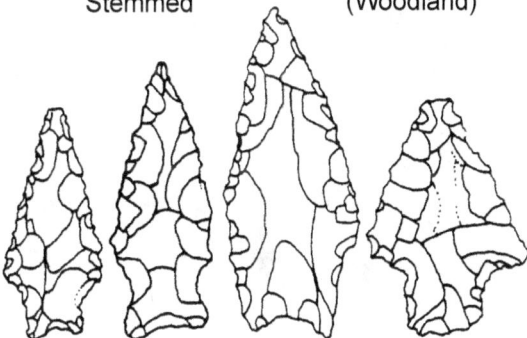

Fishtail (Archaic)   Savannah River (Archaic)

Some projectile point types of the periods of human habitation on or near the Chessie Trail.

*During early spring, when the river is high,
water fills the old channel of Reid's Lock.*

The Canal and the Chessie Trail
D.E. Brady, Jr.
Updated by Tom Kastner 2009

In the late 1700s and early 1800s the Maury River—
then called the North River—was an important artery of
commerce for the Rockbridge County part of the Valley
of Virginia. The area was predominantly agricultural,
but it included a budding iron industry. To develop, it
needed dependable and economical transportation of its
products to market, and of manufactured goods to the
farmers and householders. In spite of their present-day
rough and inadequate appearance, the James and North
rivers were much used for that purpose.

First there were the batteaux. These large boats,
roughly 45 by 7 feet, were assembled and loaded at
boatyards. They were held there until there was enough
flood water for them to be floated downstream to

Lynchburg or Richmond. Lynchburg was an important trading center for the people of what was then, pre-Civil War, central Virginia, including Rockbridge; Richmond had ironworks and flour mills, and being at the head of ship navigation on the James, it was a point from which goods could be shipped overseas. At the destination cities, some of the batteaux would be sold for the lumber from which they were built; others brought manufactured products back upstream, laboriously poled against the current by their crews. In order to facilitate the operations, river improvements were made: debris removed, rock obstructions blasted, and diversion dams and sluice-ways constructed. Wagons could and did make the trip to Lynchburg or Richmond, but in the absence of good roads, boats were more efficient in moving heavy loads.

George Washington and others envisioned a canal system up the James, over the mountains, and into the Kanawha River system (through what is now West Virginia) to the Ohio country. The Chesapeake Bay would thus be connected to the Ohio River, and that valley and the area between would be opened to Atlantic trade. Since the lower Mississippi was controlled by the Spanish or French until 1803, it was not considered a viable outlet for the Ohio country.

The James River and Kanawha Canal was eventually built as far upstream as Buchanan, in what is now Botetourt County. Considerable construction was done farther up the river, but the canal was not operated beyond Buchanan.

As it became apparent that the canal from Richmond to Balcony Falls (at what is now Glasgow) would be a reality, the North River Navigation Company was formed for the purpose of building and operating a canal from Balcony Falls to Lexington. There was at first enthusiastic support, then financial trouble. Before it had completed its work, the North River Navigation Company had to be taken over and the canal finished by the James River and Kanawha Canal Company.

An important and successfully implemented decision was to start construction at Balcony Falls and continue to work up the North River, opening sections for use as they were completed. The ends of these sections became important terminals. Two of them were Miller's Landing at the mouth of Buffalo Creek and Moomaw's Landing at what is now Buena Vista. The importance of Miller's Landing was probably enhanced by the fact that it had been a boatyard for batteaux before the canal arrived. There was a boatyard road from Lexington to this landing. What is now Houston Street in Lexington was a part of that road.

When the canal was completed to Lexington, it made the little town an inland port of some importance. Although the canal was not a financial success as such, it contributed greatly to the development of Lexington. Richmond had become a great wheat market. The big mills were shipping flour all over the world, and they depended on the canal for their supply of wheat. The Rockbridge area was a major wheat producer at the time, and was also a producer of iron through the Civil War. Much of the iron also traveled by canal to Lynchburg and Richmond. In turn, the canal made it much easier to bring manufactured items into the county, so local residents enjoyed a better supply of goods than previously.

In November 1860 the first canal boat reached Lexington. An 1866 issue of the *Gazette Banner* carried an advertisement by some local farmers soliciting business for the canal boat *E. L. Chirm,* captained by Henry W. Guy. They operated the boat to provide "lower freight rates" for farmers. The same issue has an ad by A. Alexander and Company, a general mercantile business "on the canal," for an express freight service between Lexington and Richmond.

Packet boats full of passengers also plied the canal. In order to fit into the locks, a boat was limited to 95 feet in length, but it had a kitchen, bar, lounge, toilet, and sleeping facilities. It did not have exactly "stateroom" accommodations; at bedtime, mattresses were put on the tables and a curtain was drawn to separate sleeping areas for men and women.

Miley photograph, courtesy of W&L Special Collections and Sally Mann

A canal boat in the Lexington Basin, (ca. 1865),
is at approximately the location of the
present US 11 bridge. In the background are
Stono mansion (left) and the ruins of VMI (right),
which was burned by Union Gen. David Hunter in 1864.

The boats were designed to accommodate 60 people. Trips were frequently social events, always accompanied by a banjo or fiddle player. At Jordan's Point in Lexington, stone wharves with mooring rings can still be seen, as well as the foundations of buildings that recall the canal in its heyday. Surrounding the Point was the Lexington Basin, a stretch of water that was the canal equivalent of a railroad yard.

The canal used water from the river but did not necessarily follow the stream channel. The canals were ditches wide and deep enough to accommodate the boats. These ditches were filled with standing or slowly moving water. Alongside the ditch was a towpath on which horses or mules walked to pull the boats. The pace was slow, and as a result the waves produced by the movement of the boat were minimal and there was little erosion of the earthen sides of the canal. Steam-powered boats were tried near Richmond but caused so much erosion that they could not be used.

Miley photograph, courtesy of W&L Special Collections and Sally Mann

*Reid's Lock, shortly after
construction of the canal system.*

As boats were going upstream, the waters in the canal would have to be raised to a higher level. This was accomplished by means of locks. A stone and timber structure, a lock typically formed an enclosure 15 ½ feet wide, 100 feet long, and as high as was required to reach the new elevation. At the ends of the enclosure were heavy wooden gates made as nearly waterproof as possible. A boat going upstream entered the lock with the gate at the upper end closed, holding back the water above. Once the boat was inside the lock, the downstream gate was closed and a small wicket gate or valve in the upper gate was opened, allowing the water to enter and fill the enclosure. As the water rose, it carried the boat to the level of the upstream end. The large upper gate was then opened and the boat was propelled out of the lock on the new level. To go downstream, the operation was reversed.

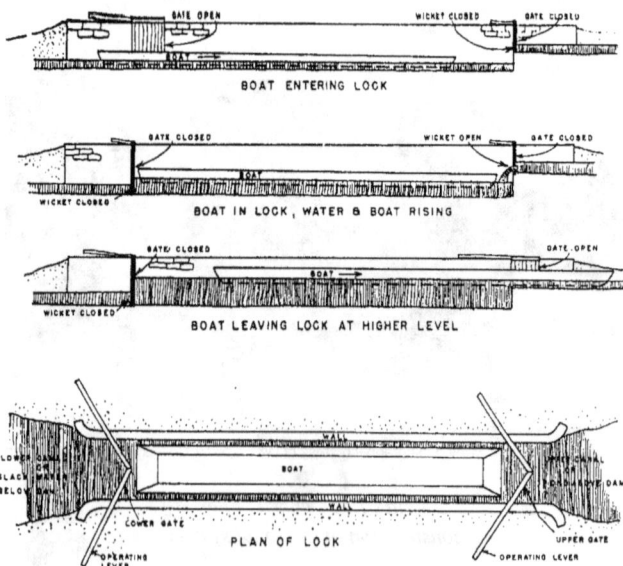

A diagram of typical lock operation in the
North River Navigation canal system.

The locks along the Chessie Trail are masterpieces of the stoneworker's art. Large, intricately shaped stones were fitted together to make grooves and notches for the wooden gates. The stones in the Balcony Falls lock have "marks" inscribed on them to identify the stonecutter. For some reason no such marks appear on the stones along the Chessie Trail.

Near each lock was the lock keeper's home, usually a rather crude cabin. None of these houses still exist along the Chessie Trail. The crew of a boat would sound a horn to alert the lock keeper of its approach, and he would make ready to receive the boat.

All told, the canal system consisted of some 10 miles of ditch with 14 locks, plus 10 miles of slack water behind 10 dams. These provided a lift of 188 feet from the James River to Lexington

From Glasgow to Buena Vista there was extensive use of dug canals, locks, and dams. From Buena Vista to Lexington there was only one-half mile of dug canal, at South River. The rest of the distance was covered by slack water from one dam to the next, with a lock at each of four dams along this stretch to lift the boats.

The towpaths ran along the edges of the slack-water ponds formed by the dams. When the terrain made it necessary to change sides of the river with the towpath, the tow animals were ferried across to the other side.

Railroad builders found the towpaths tempting places for laying track, but not always structurally strong enough to support trains. There was considerable trouble with subsidence and collapse.

The Chessie Trail in some places is actually on the original towpath, and throughout the length of the Trail, its remnants can be seen. About a mile below the Mill Creek gate, near the mouth of Huffmans Run, old canal boats were apparently tied up and abandoned. Over the years, as the water has shifted the sediment, fragments of these boats have occasionally been seen.

Reid's Dam, just below old milepost 2, formed the slack water for the Lexington Basin. The lock at Reid's Dam also made the transition to the slack-water pond behind the South River Dam. After it was no longer used for the canal, Reid's Dam served as a source of hydroelectric power until sometime before the Great Depression. Studies were occasionally made thereafter on the feasibility of resuming power generation at the dam.

Photo by Patte Wood

The remains of Reid's Dam today reflect the damage wrought by Hurricane Camille in 1969. The dam was one of a series that created slack-water ponds navigated by canal boats in their progress up and down the river.

The South River Dam, at old milepost 4, was blasted in order to reduce flooding in the years after the canal ceased operation. The main lock was on the south side, but on the Trail side are visible the remains of the foundations of a mill, which used water from the dam, and a partial lock. The partial lock suggests that a canal might have been planned for South River. At the community of Mountain View just up the South River from the old dam site were a mill, a furnace (its remains still visible on South River Road), and a marl mine. Products from these would have been shipped by the Maury (or North River) canal, and it is possible the canal planners envisioned a branch line to this point.

The South River dam was a half-mile upstream from the slack water behind the Ben Salem Dam. The gap was filled by a section of canal with locks.

Ben Salem Dam was located near old milepost 6. Its remains are rocks in the river and the well-preserved lock on the south side of the river, across from the Trail. Upstream from the lock is a clearly defined portion of the towpath. The Ben Salem Lock lowered boats into the slack water of Zimmerman's Dam, which was one mile downstream. A small part of this dam and its lock are in fair condition and may be seen from the Trail. Since this lock is on the north side, there was a tow path cross-over between it and the Ben Salem Dam and lock.

Zimmerman's Dam was used for hydroelectric power generation and was the source of electricity for the then-new city of Buena Vista, founded in the late 1880s; at that time it was known as "Lighthouse Dam." Zimmerman's Lock discharged into the slack water of Moomaw's Dam, which—with its lock—was later incorporated into the Georgia-Bonded Fibers factory (now Bontex) at Buena Vista.

Although the canal operated about 20 years, the seeds of its demise had already been sown by the time it reached Lexington. By 1860 there was a railroad operating through Goshen in the northwestern part of the county. Claudius Crozet was promoting the railroad as the better means of transportation. The canals required considerable maintenance and were often unusable for some time following spring floods. By 1880, the high costs of operation and the increasing capabilities of rail transportation resulted in the sale of the James River and Kanawha Co. to the Richmond and Alleghany Railroad. Tracks were laid along the tow paths and canal operations ceased. What had been a canal from the James to Lexington became a single-line spur of a railroad following much the same path.

How this new mode of transportation flourished and was in turn replaced, with its footprint becoming the Chessie Trail, is yet another of history's many tales. ❧

References Cited

Some of the data for this chapter were obtained from the following sources:

Couper, William, "Claudius Crozet," *Southern Sketches*, Series 5.

Gilliam, Catherine M., "Jordan's Point—Lexington, Virginia: A Site History," *Rockbridge Historical Society Proceedings* 9 (1975-79): 109-38.

Knapp, John W., "Trade and Transportation in Rockbridge: The First Hundred Years," *Rockbridge Historical Society Proceedings* 9 (1975-79): 211-31.

Trout., William E., III, "An Automobile Tour and Field Guide to the North River Navigation" (Lexington: Rockbridge Historical Society, 1983). A pamphlet.

The C&O en route to Lexington sometime in the 1880s.

The Railroads and the Chessie Trail
Matthew W. Paxton, Jr.
Updated 2009

The Chessie Trail is a legacy of the wheeling-and-dealing, boom-and-bust post-Civil War era, when the railroads dominated the news. Even quiet little Lexington, Virginia, felt great ambitions to become a railroad center. In the local news sheets the railroads vied for top billing with the statewide political battle between the "funders" and the "readjusters."

At that time, the town's railroad ambitions were not so farfetched, for in 1881 the village of Big Lick, fifty miles down the Valley, had a population of but 400 souls. In that year, east-west and north-south rail connections were secured for that town, which was soon to become the railroad boom city of Roanoke.

Also in 1881, the first train from Richmond chugged into East Lexington over what is now the Chessie Trail, and Lexington achieved its long-awaited rail link with the outside world.

The years 1880-81 were pivotal ones for railroading in Rockbridge. This period saw not only the completion of the first railroad into Lexington, but also the proposal of a grand scheme to connect Pittsburgh with Lynchburg by way of Lexington and Goshen Pass, as well as the announcement of plans to complete the long-delayed Valley Railroad from Staunton to Salem through Lexington.

Finally, these years saw the beginning of the successful "end run" of the Shenandoah Valley railroad from Waynesboro to Roanoke, through the lightly populated eastern edge of Rockbridge and what later became Buena Vista. This road would eventually become part of the Norfolk and Western system.

The *Lexington Gazette and Citizen* reported on March 11, 1880, that "the new railroad up the James River Valley is assured, and it gives promise of a new era for Virginia and especially for our section, so long locked up, with its rich mines of minerals undeveloped and its broad fields without an outlet to market." The newspaper assured its readers that the Lexington branch of the new railroad "is as much a part of the road as any other section of it."

This first railroad into Lexington had its beginning when a bill authorizing the sale of the James River and Kanawha Company's canal works to the Richmond and Alleghany Railroad became law in 1878. This move followed the disastrous flood of 1877, which dealt the canal a severe blow. The rail line was to follow, basically, the old canal towpath.

The final contract between the canal and railroad companies was executed March 5, 1880. It provided, among other things, that the railroad was to have the same gauge as the Chesapeake and Ohio Railroad (i.e., the railroad standard of 4 feet, 8½ inches), that construction was to be done so as not to interrupt the business of the canal in those sections not occupied by the railroad, and that when the railroad supplanted the canal it should pay the owners of all canal boats a fair cash value for their boats.

Track laying began July 8, 1880, at Maiden's Adventure, near Richmond, and on Friday afternoon, October 14, 1881, the last rail was laid on the Lexington division, the main line from Richmond to Clifton Forge having been completed a month earlier. The next day trains started almost simultaneously from Natural Bridge, Lexington, and Richmond, meeting at Howardsville, where the road was formally declared open.

photograph courtesy of W&L Special Collections

The C&O tracks just east of the US Route 11 highway bridge, which was built in 1935. This photo is probably from the 1940s.

The *Lexington Gazette and Citizen* reported that on the eve of the official opening, "the special train reached Natural Bridge about sunset and the official party there met President French [of the Richmond and Alleghany] and his party who had come from New York and Boston via the White Sulphur."

A banquet was held at The Bridge and during the evening's toasts one speaker commented that "Richmond, the queen city of Virginia and the queen of the James River Valley, would soon be wedded to the Great Lakes by the Richmond and Alleghany Railroad." In obvious reference to the Northern capitalists involved in the railroad, a toast was raised to the union of the states. The union was compared to the Natural Bridge, where "we once thought there was a chasm between the north buttress and the south buttress, but if we look high enough we find they are joined together."

The arrival of the railroad, bringing fast, modern travel to Lexington, was heralded by the local newspaper with the exclamation, "hereafter, the whistle and not the horn [of the canal boat] will announce the arrival and departure of travelers to and from Lexington."

But the advent of rail transportation turned out not to be an unmixed blessing. It was not long before the *Gazette and Citizen* reported: "We regret to say that the Richmond and Alleghany Railroad again desecrated the Sabbath day and caused others to desecrate it, in running an excursion train from Lynchburg to this place on Sunday last. We hear of conduct upon the part of certain persons who came upon the train which a proper regard for our readers compels us to refrain from publishing. We would be glad, if our town authorities would take such steps as would prevent our town being deluged with a class of people detrimental to decency and good order on the Sabbath day."

Ironically, the first railroad to arrive in Lexington was not the one that people had been pinning their hopes on. The rail connection most desired was the one with Baltimore. In October 1866 the citizens of Rockbridge

County approved a $100,000 bond issue to help finance the Valley Railroad. The company would construct a line from Salem to Harrisonburg, where it would connect with other lines reaching up the Valley to Harpers Ferry, at which point it would connect with the Baltimore and Ohio system. Additional county bond issues were voted by overwhelming majorities in 1868 and 1871. The total Rockbridge subscription amounted to $525,000.

Work on the line commenced in earnest and it was opened from Harrisonburg to Staunton March 3, 1874, but the effects of the financial panic of 1873 forced suspension of construction in December 1874. The long delay that followed probably sealed the doom of Lexington as a railroad town.

At the time the Richmond and Alleghany Railroad reached Lexington with its line up the James and North (Maury) rivers in 1881, work on the Valley Railroad was moving forward at a painfully slow pace. The lead article in the *Gazette and Citizen* on October 18, 1883, was headed "Almost Here," and began, "For 10 years we have been talking and writing about the Valley Railroad, alternating hopes and fears finding expression in what we had to say .... We are gratified to say that the Lexington connection is at last virtually accomplished. The track on Monday evening was just opposite Col. Harman's house and about 500 yards from the trestle across the Valley Pike at Willow Springs, leaving a gap of 1¼ miles, which the manager informed us would be completed by Thursday evening, and before our country readers receive the Gazette this week, the connection will be made with the Richmond and Alleghany Railroad at the River. The first train through from Staunton will bring the material for the construction of the Lexington Depot, which will be of Baltimore pressed brick and will be one of the handsomnest [sic] on the line of the road."

photograph courtesy of W&L Special Collections

*Platform of the old Lexington rail station, which operated at its
original location at the foot of McLaughlin Street until the late 1960s,
though with diminishing traffic. At left is the old Higgins & Irvine
lumber yard building, now site of Lexington Building Supply and Mill.*

The next week the paper trumpeted "Finished," and
told that "the first through passenger train on the Valley
Railroad (a special), arrived at the junction with the
Richmond and Alleghany Railroad on Saturday evening
last, with the cadets of the VMI returning from the
Winchester Fair."

The two rail lines were connected with a "Y" for
turning the trains around. The R&A granted the Valley
Railroad trackage rights from the "Y" to its east terminus
and, in turn, the Valley Line granted the R&A trackage
rights from that point to the station in Lexington. Turning
at the "Y," trains had to back up a steep grade into
Lexington. Schedules published in the local newspapers
show that by 1885 each rail line had two trains coming
into and going out of Lexington daily.

The Richmond and Alleghany, which had carried a heavy debt since its beginning, was taken over by receivers in 1883, and its visions of pushing on to the Ohio River were abandoned. It was again reorganized in a "friendly foreclosure" in 1885.

The collapse of a wooden trestle over South River on the Lexington Branch in the fall of 1885 brought forth an angry blast at the substandard condition of the railroad. An anonymous "citizen" wrote in the *Rockbridge County News* of November 6, 1885: "As everybody knows the Richmond and Alleghany Railroad, especially the Lexington Branch, was not built at all; it was just kicked together.

Photo by Patte Wood

Remains of the Chessie Trail footbridge that crossed the South River atop the old C&O trestle. The footbridge was destroyed by a flash flood that claimed several lives along the South River in 2003. The pilings seen in this 2009 photo were part of the old railroad trestle.

"Pine trestles were set upon mud sills, or piles, across deep ravines and dangerous streams, such as North and South rivers, and long bridges were balanced on the slender rock piers built for the mule walk or towpath at the old aqueducts." The writer cited the urgent necessity for some effective legislation, saying "the railroad companies should be compelled to erect good iron bridges over dangerous streams."

The redoubtable Colonel Henry Clay Parsons, a key figure in the formation of the Richmond and Alleghany Railroad, was one of eight passengers aboard the train that plunged into the flood waters of South River that October night when the trestle gave way. In a letter to the general manager of the R&A he described the wreck: "Suddenly the rear of our coach (we were next to the engine, the baggage car behind) was lifted. The car seemed to hang in the air and shook violently.... The car dropped heavilly [sic] then swung to the left and lurched to the right, throwing the passengers to that side. I shouted 'hold to your seats.' The front of the car fell into the water and the car lurched back to the left. The water broke through the lower windows and the lights went out. I succeeded in reaching the end of the car and found the door fast but was able to wrench it open. The passengers climbed up by the seats and were helped out." The engine had disappeared into the swollen waters of the river and the engineer and firemen were drowned.

The condition of the railroad and its vulnerability to high water continued to be a local concern. The *Rockbridge County News* reported on June 6, 1889, that in a flood that was not as bad as those of 1870 and 1877 "the Richmond and Alleghany Railroad was badly torn up along the James River and above Miller's on North River." And in August of that year the newspaper noted: "The railroad hands are thoroughly overhauling and replacing with new timbers much of the old trestles at the river at this place. This is well enough and a timely precaution after the continued high waters of the summer, but it is about time this long stretch of wood trestles give place to a more substantial bridge."

In view of the shaky condition of the R&A, general satisfaction was expressed when it was taken over by the Chesapeake and Ohio early in 1890 through a foreclosure sale. *The Richmond Times* commented, "the benefit to the State of Virginia by this affiance can scarcely be overestimated."

The C&O had just gained a direct connection with Cincinnati and was riding high. In the decade between 1880 and 1890 the C&O more than doubled its trackage from 436 to 953 miles. Meanwhile, the Norfolk and Western Railroad made comparable gains in the length of its system.

Times were booming in 1890 and the people of Lexington could not imagine that their town would be left at the dead end of two branch lines. In September 1889, the stockholders of the Pittsburgh and Virginia Railroad had met in Lexington and elected officers. "It is proposed," reported the *Rockbridge County News*, "to run a direct road from this region through to Pittsburg[h], Pa." The extension of the Valley Railroad to Salem was also being discussed. Much of the grading and culvert work for the extension had already been done.

At the end of 1889 the *Rockbridge County News* was reporting that "Buena Vista is enjoying a live boom…. The value of lots has jumped about 100 percent in the past two weeks."

At the outset of 1890 the *Rockbridge County News* was beating the drums for the new western rail connection. Wrote the editor: "With this new road with connections from Pittsburgh in the West to the Virginia ports on the sea; with its great trains lumbering down through Goshen Pass; on down North River to Lexington, and thence across country to Glasgow, what new development shall we see in this already booming county of Rockbridge. What then for that 'finished town' of Lexington? True, there are about her many dear associations and historic memories, but don't you forget it, she is coming, Brother Jonathan, she's coming. Let's all lend a hand."

Photo by Patte Wood

Boxerwood education director Elise Sheffield inspects the old "BF 16" marker, which told C&O engineers that they were 16 miles from Balcony Falls.

The local paper was also pushing for the C&O to "fill the gap" between Lexington and Goshen to "facilitate communication between the main lines and make this cross-road with its fine country and mineral resources a feeder to both." In calling for improvement of the trestle at Lexington, the newspaper declared of Lexington, "this is an important point, and will grow in importance with the extension of the roads."

Much of life in Lexington revolved around the arrival and departure of the trains. In his book *Only Yesterday in Lexington, Virginia*, General John Letcher recalls that prior to World War I there were 12 passenger trains each day in the Lexington station, six arriving and six leaving. And there were two freight trains daily.

The trains on the Valley line, which had become a branch of the B&O, departed at 8 a.m., noon, and 3 p.m. and arrived at 10 a.m., 2 p.m., and 6 p.m. These local trains took an hour and forty minutes to reach Staunton where connections could be made with the main line of the C&O, or passengers could continue up the valley by way of Harrisonburg to Washington. Stations on the B&O in Rockbridge were at East Lexington, Timber Ridge, Davis, Fairfield, Decatur, and Raphine.

The C&O trains left Lexington at 7 a.m., 11 a.m., and 4:30 p.m. and arrived in Lexington at 9 a.m., 3:30 p.m., and 6:30 p.m. At Buena Vista they connected with the main line of the Norfolk and Western for north- or south-bound trains and at Balcony Falls with the main line of the C&O for east- and west-bound trains. It took the local trains about an hour to make the 21-mile trip from Lexington to Balcony Falls. They made stops at East Lexington, South River, Buena Vista, and Buffalo Forge.

On July 1, 1903, after much debate about tax loss to the county, the C&O ceased to operate its tracks between Balcony Falls and Loch Laird, just below Buena Vista, and entered into a lease with the N&W for joint trackage between these two points.

Letcher recalls that the B&O passenger trains generally had three cars, two for passengers and one for mail and baggage. The C&O local usually had only two cars, one to accommodate passengers and one for mail and baggage. Local rail service reached its zenith, notes Letcher, just prior to World War I when the B&O added a parlor car with individual seats and a kitchenette on a train that went all the way to Washington. At that time the departure of the morning train was moved back from 8 to 4 a.m.

The big event of the day was the arrival of the evening train from Washington. Letcher recalls that it "came backing into the station with its brightly lighted resplendent parlor car just as darkness was settling down on the town in the winter evening." At that time all the cabs in town and the special horse-drawn vans of the Lexington Hotel and the Central Hotel would converge on the station.

Financial reports from the early years of the Richmond and Alleghany show that nearly half of the line's passenger traffic came from chartered excursions during the spring and summer months. For a local resident recalling those excursions, the things that stood out were lunches packed in shoe boxes and eyes full of cinders.

In Lexington, railroading was not without its lighter side. Among the most important special trains from Lexington were those carrying the W&L students and VMI cadets to football games. Letcher recalls that if the Washington and Lee football team was returning from a game on the Sunday evening train, VMI cadets would slip out of barracks and grease the rails of the track at the steepest point on the grade behind VMI. When the train backed in, the wheels of the engine would spin on the greased rails and the train would come to a halt. After several fruitless efforts to get up to the Lexington station, the train would slowly return to East Lexington where the team would disembark. Meanwhile, the welcoming student delegation would wait in vain at the Lexington station to welcome the team home.

By the 1920s, the automobile was rapidly supplanting the railroad as the primary means of transportation.

The B&O's line from Staunton to Lexington was abandoned in 1942 and the rails were melted down for the war effort. Passenger service on the C&O's line from Balcony Falls to Lexington quietly ended June 30, 1954, although it was announced that the railroad would continue to provide Lexington with freight service six days a week.

A reporter for the *Rockbridge County News* rode on the last train carrying passengers from Lexington to Buena Vista. He noted that the train, pulled by a diesel engine, included several freight cars, a wooden passenger-baggage car and a caboose. The only other people aboard the train were the brakemen and the grandson of the conductor.

The train left Lexington at 12:05 p.m. It took 25 minutes to get to the outskirts of Buena Vista. But three stops to switch cars delayed its arrival at the station there until 1:10 p.m. After the hour-long trip the reporter disembarked, and the train went on its way to Balcony Falls.

On August 20, 1969, rail service to Lexington was abruptly ended when the floodwaters from Hurricane Camille ruined the C&O's line along the Maury River and destroyed the wooden trestle at East Lexington. On September 1, 1970, the Interstate Commerce Commission issued an order granting the C&O permission to discontinue service on the branch from mile post 12, under the US 60 bridge at Buena Vista, to the end of the line at mile post 21 in Lexington.

The railroad had requested the abandonment because of the high cost of replacing the line and because the line had become a deficit operation. Its volume at the time of the Camille flood was less than a carload a day. Opposition to the abandonment, initially discussed by the Lexington City Council, had been dropped. With the ICC action, Lexington was officially without rail service for the first time in nearly 90 years. ❦

The old Tankersley's Tavern, now a private home, stands at what was once the northern end of a covered bridge across the Maury.

Historic Structures Along the Chessie Trail
Royster Lyle, Jr.
Updated by Eliot Balazs 2009

Introduction to the 2009 revision

To walk the Chessie Trail is to walk through history. The hiker is greeted by an unfolding historical tableau offering the remains and relics of eras gone by. The original rail bed and its companion, the Maury River (once called North River), wind their way from Lexington to Buena Vista in a symbiotic journey accompanied by the remnants of structures once vital to Maury River navigation and to the railroad system that served the communities of Rockbridge County. Once in time packet barges and batteau boats, laden with their material and human cargo, plied the lazy waters of the Maury. Once

177

in time thundering steam locomotives towed their vital loads across now absent rails. These systems of river and rail traffic were life-supporting arteries to the people they served. Now, although the locomotives are silent and long since departed, and the river hosts canoes of sportspeople instead of canal boats of cargo, the physical remains of major engineering efforts to tame the river and support the rails once so vital to the area still exist. The Chessie hiker may see the remains of lock systems, dams, bridges and other critical structures supporting canal and railroad operation. Massive stonework for the crucial canal lock systems, today covered with natural growth, still stands for the ages. Concrete abutments, although no longer supporting their bridge girders, still stand where placed by the hands of those who labored decades ago. Here and there even a steel rail, forgotten in the haste of railroad removal crews, lies nestled in the undergrowth by the side of the trail. Observant hikers will be greeted by all these things and more on their journey along the Chessie Trail, a journey back in time.

—*Eliot Balazs, 2009*

Jordan's Point, officially still the head of the Chessie Trail, was once the site of much business activity. In the canal days there were wharves and warehouses here; later there were mills and manufacturing enterprises. Several severe floods and changes in the highway and the railroad have made a great difference. Today, Jordan's Point is the site of Jordan's Point Park, a large attractive green area located on an island (traditionally known as VMI Island) and peninsula formed by the Maury River, Woods Creek, and an old mill race. It is a 9-acre facility with picnic tables, walking trails, and canoe launch. It is currently being developed in stages and most recently was designated as a point on Virginia's Civil War Trails program for visitors to observe where Union General David Hunter crossed into Lexington after shelling it in June of 1864.

Jordan's Point Park was also at the southern end of the Chessie Trail footbridge across the river, destroyed by flooding on two occasions. The bridge was not rebuilt after a flood in the mid-1990s, and hikers now often pick up the Trail about a mile downstream, near Mill Creek. (For details on the Mill Creek entrance and also on how to hike the Trail from Jordan's Point to Mill Creek, see the Trail Log at the end of this guidebook.)

At the Trail's head at Jordan's Point, a brass plaque marks the official opening of the Trail in 1981. From this vantage point the hiker has a good view of the early town dam, built to force the river into the mill race, thus forming the island. An interesting feature of the present concrete dam is the fish ladder on the north side of the dam, although the ladder is in some disrepair.

Also visible from the Point are the large stone abutments of the 1870 covered bridge that survived until the 1940s. Unfortunately the abutments were hit hard by the November 1985 flood and subsequent flood events, and have begun to deteriorate badly. Around the shores of VMI Island low stone walls can still be seen, relics of the canal era of the mid-19th century.

Across the river is East Lexington, which has played an important part in the history of the community. Here the Valley Road crossed the river, first by means of a ford and later on the covered bridge. From 1818 on, John Jordan's manor house, Stono, looked down on the activity on the river below, and it was during the heyday of river traffic that the area along the waterfront became known as Jordan's Point. For hikers who wish to continue to explore after a day on the Trail, a picnic in the park may be followed by a walk past the old miller's house in the park and on up the steep, winding road at the back of VMI. At the crest, Stono still stands, and may be seen on the way into the VMI campus itself.

When the canal from Richmond finally reached Lexington in 1860, Jordan's Point became the terminus for the system that Jordan himself had built. During the latter part of the 19th century, the area on the south side of the river was called Beehenbrook, and the north side Lavesia.

The first covered bridge was built in 1835 by Jordan, who was allowed to collect tolls from the users. It was burned in 1864 by Southern troops retreating from Union General Hunter's invasion.

After the war there were several attempts to get a new bridge. Finally in 1870 the abutments were raised and a single-span covered bridge was built by John Wood, who built many wooden bridges in the Valley during this period. After the new Maury River bridge on Route 11 was completed in 1935, the 1870 covered bridge fell into disrepair, and in spite of a number of efforts to save it, the bridge was finally declared unsafe in the 1940s and taken down.

On the north side of the river, six houses are visible from Jordan's Point, sitting in a row at varying angles to the old highway, now County Route 631 or Furr's Mill Road. Of these buildings, which once looked down on the canal basin, old Tankersley's Tavern, or Old Bridge, is perhaps the best known. The tavern was built in various stages, and the date of the earliest portion is still unknown; it is believed to have been built by Samuel F. Jordan, son of John Jordan.

One of the 20th century owners of the tavern recalled that William Netz, an agent for the canal line, operated a bar there before the property came into the Tankersley family in 1886. Frank Tankersley recalled in 1946 that the large rambling building that stood just at the northern end of the covered bridge was a "favorite spot for refreshments among salesmen and travelers on the canal packet boats, which moored at the nearby docks." In the 1950s a portion of the building, its east side, was taken down and the remaining section turned into an attractive dwelling.

Among the other interesting features in East Lexington that can be seen from the Trail is an early brick beehive cistern (to the left of the frame house located where the end of the footbridge used to be). On the north side of the river, just below the East Lexington houses, the Trail goes under the 1935 Route 11 bridge and begins a long straight stretch behind a number of 20th-century warehouses and the old Pure Oil site.

Just before the Trail crosses Mill Creek, a large manor house can be seen to the left, across and above County Route 631, which on this side of the river is known as Old Buena Vista Road. Clifton overlooks the river and the high cliffs where the Maury begins to make a large horseshoe bend. The property where Clifton now stands originally belonged to William Alexander, who ran a business in the area about 1778. It seems likely that his son, Major John Alexander (1776-1853), built this fine country house in the early part of the 19th century.

Photo by Patte Wood

Clifton, a private home, was built by the Alexander family in the early 1800s. The city of Clifton Forge was in all likelihood named for the family home, which stands on the north side of Route 631 overlooking the Trail and river.

Around 1850, Major Alexander's son, William Lyle Alexander, took over the management of a number of his father's iron furnaces and named one of them – some 35 miles to the west, near the headwaters of the James River – Clifton Forge.

The property has seen a number of changes over the years. Before the advent of the railroad and the county road, for instance, the front lawn of Clifton extended down to the river. In the early 1980s the Classical portico was added —a facade much in keeping with the tradition of Stono and several other early 19th-century Valley homes in the Lexington area.

Beyond Clifton, the Trail begins a long, dramatic turn to the right under a cliff quite obviously created during blasting for the railroad bed. The hiker can still see the clear marks of the dynamite drills along the cliff face, as well as boulders that were actually blown into the river.

About one mile beyond the high cliffs, the hiker comes to Reid's Dam with its well-preserved lock just to the right of the Trail. Here one can safely stand behind a guard rail and look down into the lock, which still has water running through it if the river level is right. The hiker can also get a close look at some of the special configurations of the lock—such as the recesses in the massive stonework that held the upstream gates of the lock.

Photo by Patte Wood

The break in Reid's Dam allows easy passage
for canoeists and kayakers on the river.

The stone and concrete dam, originally built in about 1858, survived mostly intact until the August 1969 flood, caused by Hurricane Camille, opened a wide gap in the middle. Subsequent floods, particularly one during the November 1985 storm, have taken their toll, so that now canoeists can easily paddle right through the middle of the dam, even in low water.

The Reid's Dam lock is one of the best-preserved

features of the canal system. A wide limestone shelf at the river's edge about fifty yards downstream from Reid's Lock is a good picnic spot with a clear view of the massive structure.

Photo by Peggy Dyson-Cobb

Reid's Lock, seen from above during spring flooding. Near the top of the stonework, details of the grooves and notches that held the wooden gates may still be seen.

A half-mile farther downstream, Interstate 81 passes overhead on huge pylons; here the Trail skirts bottom lands that have been used for hayfields and pasture. A short distance farther along, high cliffs come into view on the right side of the river.

South River Lock and the remains of its dam are

soon visible to the hiker, both across the river and next to the Trail. At this point, boats left the river in a specially constructed canal to bypass the confluence of the North (now Maury) and South rivers. A half-mile long, this canal has since been filled in, but the two lift locks that lowered the boats back into the river can still be seen.

Photo by Patte Wood

At the old spring house just downstream of the South River Lock, two eras of construction can be seen: the old limestone masonry at right, and 20th-century repairs to the left.

The hiker next encounters the remains of the South River Bridge, which was a 235-foot long footbridge built on a C & O trestle dating back to 1901. Oddly enough, the C&O bridge was moved from its first location in Taylor, Ky., to the South River site in 1914. The footbridge was an attractive architectural feature of the Trail until flooding in 2003 destroyed it.

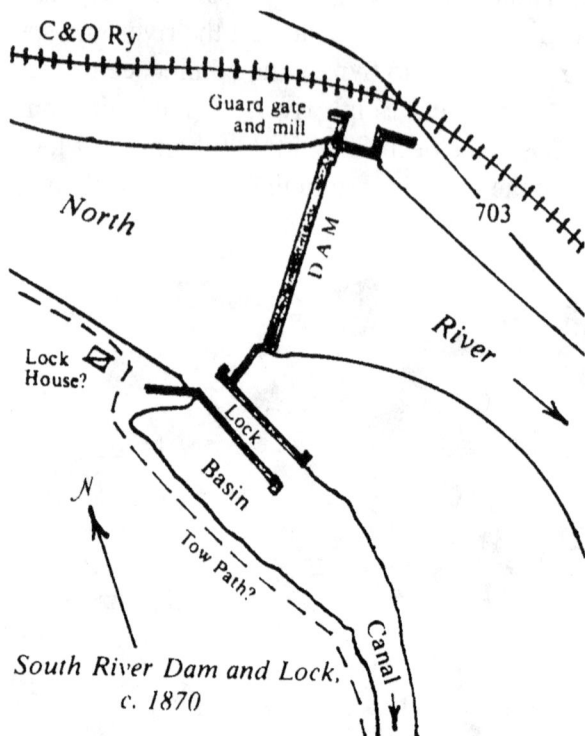

South River Dam and Lock,
c. 1870

The early 19th century country estate house, Tuscan Villa, also known as the Old Glasgow Farm, can be seen from the middle of the highway bridge looking up South River. This good example of a Valley house built in the 1820s was given a portico with large columns in the 1850s. For generations it was the seat of one branch of the Glasgow family, who were among the earliest settlers of Rockbridge County.

As of this printing, the absence of the South River footbridge, plus a washed-out culvert about a half-mile farther on, make this portion of the Trail difficult to navigate. It is anticipated that both breaks will be repaired, but at present, the best access to the remainder of the Trail is from the Buena Vista end. In anticipation of the Trail's repair, the narrative of neighboring structures continues here in order, from west – where the hiker earlier began – to east.

Beyond the bridge and on the far side of the river is an early manor house known as Hidden House. It is an 1850 Greek Revival brick house with a handsome double portico in front and two-story porches along the left side. During the canal era, the house lay between the canal and the river, and boats passed at the rear of the house.

A little farther along the Trail, a "W" on a four-foot-high concrete post once reminded the engineer on the rail tracks coming from Buena Vista to blow his whistle (long, long, short, and a long blast) upon approaching the old public road that is now County Road 839.

Down the Trail about a mile, the hiker begins to see Ben Salem Lock and the remains of its dam, which are about six miles below Jordan's Point. Although the dam is now only an irregular scattering of rocks across the river, the lock is still in good condition, and the State Highway Department has developed a very attractive picnic area on the south side of the river. It is readily available from Route 60, midway between Lexington and Buena Vista and only a mile or so from the junction of Route 60 and Interstate 81.

The Trail then takes the hiker through a long open area of pastureland, from which Zimmerman's Lock may be seen, about three quarters of a mile below the Ben Salem Lock. Hikers are reminded at this point that wherever open fields abut the trail, they are private property and are often used for pasture – a principal reason why dogs may only be admitted to the Trail on leash. It is particularly important that they be leashed and kept to the Trail where livestock are present.

Photo by Ed Spencer

Zimmerman's Lock may be seen across a stretch of pasture, about three-quarters of a mile below the Ben Salem Lock.

From here, as the Trail continues to make its way along the north bank, interesting caves left by blasting for the railroad can be seen.

The end of the Trail comes with the approach to the Russell Robey Bridge, constructed in 1984, which carries Route 60 over the Maury River. One of the considerations in the design of the new bridge was a walking/ jogging element that would allow hikers to continue from the Chessie Trail, cross the river and pick up a connecting trail to take hikers all the way to the Glen Maury Park area. This would have completed a special dream that many local people have had for the Trail from the outset, but which has yet to be realized. It is hoped that in the not-too-distant future, this connection may be the next step in the proposed Brushy Blue Trail connecting the Appalachian Trail along the Blue Ridge Mountains with trails in the Alleghenies to the west. ✹

Bibliography

Royster Lyle, Jr., Sally Mann, and Pamela Hemenway
  Simpson, *The Architecture of Historic Lexington*
  (Charlottesville: University Press of Virginia, 1977).

The Chessie trailhead plaque commemorates the official opening of the Trail, Nov. 9, 1981.

Trail Log

Edgar W. Spencer and Larry I. Bland
Revised and Updated by Edgar W. Spencer and Lisa Tracy 2009

There are four major access points to the Chessie Nature Trail: at the beginning on Jordan's Point; on County Road 631 (Old Buena Vista Road) just east of Mill Creek; where County Road 703/608 (Stuartsburg Pike) crosses the South River; and at the eastern end of the Trail, near Buena Vista on County Road 608 about a half mile north from the eastern end of the US 60 bridge over the Maury River. There are places to park at each location. Trail mileage is counted from the Jordan's Point end, so this Log of sights along the Trail will begin there. Remember as you walk, the property 50 feet off the Trail on both sides is private.

Three sets of distance markers are located along the Trail at this time. The oldest set belonging to the original C & O railway consists of concrete posts marked "BF" that show the distance of the post from Balcony Falls, near Glasgow. A second set, 6-inch-by-6-inch posts about three feet tall with slanted tops, was erected when the Trail first opened. These posts show the mileage from Jordan's Point in one direction and from the Buena Vista entrance in the other direction. The third set, about 19 inches tall, was installed after the footbridge across the Maury River was washed out.

(*Ed. note*: Digital printing and the advent of the Internet allow updates that weren't possible when this Guide was first published. In the course of this Log, you'll find several references to "breaks" in the Chessie Trail that currently impede a continuous hike. The center section of the Trail, however, gives you several miles of beautiful hiking on level ground with a variety of microenvironments, geological features, and archaeology. As the breaks are repaired and the Trail is restored, the Friends of the Chessie Trail will be updating its Web information via the Rockbridge Area Conservation Council's website, and periodic updates will be published in subsequent press runs.)

To reach the original beginning of the Chessie Trail, take Moses Mill Road— across from the intersection of Business US 11 and Bypass US 11—to Jordan's Point. The road name is a reminder that the Trail starts at the site of the former manufacturing and commercial center of the area. Parking is available just beyond the bridge over Woods Creek and across the road from the end of the Woods Creek Trail, which is on the bed of the siding that served the Point.

The hiker may wish to detour into Jordan's Point Park, where the marker for the official head of the Chessie

Trail may be seen. The park is a good picnic and play area and is the site of the former Moses Mill. Currently in the process of renovation, the old miller's house will eventually be open for visitors. The park also is home to one of the original batteaux that plied the Maury in the early canal days, and there is a good boat access.

The footbridge across the river that originally led to the rest of the Chessie Trail was washed out during catastrophic flooding in June 1995. Many hikers, in the absence of the footbridge, also choose to bypass the first section of the Trail and go directly to what we'll be referring to as the Mill Creek gate (also known as the West gate), about six-tenths of a mile down Old Buena Vista Road, where a more wooded section of the Trail begins. If that's your choice, skip sections 1 and 2 and go directly to "Section 3: Mill Creek Gate and Beyond."

SECTION 1: TRAILHEAD AND BEGINNING AT JORDAN'S POINT

If you have parked off Route 11 and want to explore Jordan's Point and the official Chessie trailhead, orient yourself by turning your back on Route 11 and looking slightly northwest. You'll see the old miller's house to your left, and the bridge just beyond the parking area.

The bridge crosses the former millrace, which served as a power source for Moses Mill and its predecessors. Both the mill and its dam were swept away by a flood in the late 1920s. The stonework visible along Woods Creek and the millrace is a remnant of the wharves along the basin at the head of canalboat navigation on the old North River Navigation system, a branch of the James River and Kanawha Canal. Water was backed up to this point from Reid's Dam downstream.

The Chessie Trail officially begins at the 1981 dedication marker in the park. Many Trail walkers will be pleased to note that the Trail is one of the longest flat stretches in the county, and its change in elevation is only about 60 feet from end to end.

A 110-yard bridge was initially constructed in the spring of 1982 on the concrete footings that bore the former railroad trestle. A flood in June 1982 destroyed most of the bridge, but it was quickly reconstructed in its present form higher above the riverbed and largely survived the great flood of 1985, only to be completely washed out in 1995. Current plans call for improved signage and footpaths on both sides of the river to get you to the next section of the Trail.

That section starts below the Route 11 highway bridge and goes east along the river behind the old "industrial" stretch of Old Buena Vista Road. To access this part of the Trail from Jordan's Point Park on foot, retrace your path back to Route 11. Cross the highway bridge, using the sidewalk on the upstream side of the bridge. At the north end of the bridge, turn left onto Furrs Mill Road. Walk along Furrs Mill for a short stretch to a footpath that leads down the embankment to the river's small floodplain here. Then turn left again on the floodplain and take the Trail under the highway bridge and on to the short industrial section of the Trail, which runs behind old warehouses to Mill Creek.

From the old trailhead at Jordan's Point, you may be able to see remnants of the stone abutments of the covered bridge that remained in use until the highway bridge was opened in 1935. The US 11 bridge is 638 feet long, and its road surface is 30 feet above the Trail. The old covered bridge deteriorated and was torn down in 1946. One of two large houses with two-story porches (the one on the left) across the road from the north-bank abutment was formerly Tankersley's Tavern, long one of the area's most important businesses.

As you walk the "industrial" section of the Trail, after passing under the highway bridge, you are now on the old C&O railbed itself. The concrete footers in the river upstream of the bridge are all that remain of the railroad bridge that carried trains to Lexington and then provided the base for the Trail's original footbridge. Uphill to your left are storehouses and the old Pure Oil depot. Ahead of you lie Mill Creek and the ruins of its millrace.

Many of the observations needed to piece together the geologic history of the Maury River and its valley can be made along the Trail. The channel below the highway bridge is filled with gravel being transported downstream by the river, ultimately to the ocean. Note that most of this gravel is composed of quartzite and red hematitic (iron-bearing) sandstone eroded from the ridges to the west. Gravels similar to the ones in the modern stream channel are present in road cuts along the Old Buena Vista Road. These indicate that the Maury or one of its tributaries once flowed at this level more than a hundred feet above the modern channel. These iron-bearing rocks were mined for iron in many parts of the Appalachian Mountains. The closest outcrop of these rocks is at Goshen Pass. The extraction of the iron from the sandstone was an important industry in the 19th century, and the river was one of the chief methods of transporting pig iron to mills in Lynchburg and Richmond.

The most common large native tree along the river is the distinctive sycamore, considered to be the most massive tree in the East. Its hard, coarse-grained wood is used for boxes, barrels, and furniture. These trees may live more than 500 years, providing twigs for deer and muskrats to eat and cavities for nests and shelter for wood ducks, opossums, and raccoons.

SECTION 2: FROM ROUTE 11 BRIDGE TO MILL CREEK

In general, along this section walkers will notice the deleterious effects on the plant life along the Trail of prolonged human contact where industry has been the prevailing use of the land. There are relatively fewer native plants and more invasive species than you'll see on other parts of the trail. At ground level, species such as garlic mustard, vetch, honeysuckle and multiflora rose are in evidence; overhead, the box elder and non-native ailanthus (tree-of-heaven) prevail.

Photo by Peggy Dyson-Cobb

Showy Orchis blossoms, surrounded by wild geranium foliage, are among the many native wildflowers that grace the Trail, especially in spring. Please capture them by camera only.

The box elder, officially known as ashleaf maple, is generally thought of as a nuisance tree, and indeed it does edge out other species where it thrives, but it's not entirely useless. Its soft white wood is used for boxes and crates. Syrup can be made from its sap, and its seeds provide food for squirrels and birds.

Two hundred yards downstream, off to the river side, stands the first of the concrete "BF" markers for this end of the Trail. The letters stand for Balcony Falls, the point on the James River where the Lexington Branch joined the C & O railroad's main line; the numbers represent the miles to that point.

Evidence of the Trail's former purpose are still buried in and along the sides of the old railbed: ties, tie plates, spikes, and occasionally even pieces of rail. Here and there you can still discern, in the alternating light and dark stripes perpendicular to the Trail, the places where ties were once embedded.

Along this section, outcrops of the Ordovician age Edinburg Formation are almost continuously exposed in high cliffs across the river. Most of the rocks in this formation are fine-grained, black limestones that formed in moderately deep water on the continental margin about 450 million years ago. These deposits are named for exposures that occur at Edinburg, Va., but the unit and its equivalents that have other names extend hundreds of miles up and down the Valley of Virginia and Pennsylvania.

The Edinburg is about a thousand feet thick, but it is exposed along the Maury for more than a mile because it was folded when the Appalachian Mountains were formed late in the Paleozoic Era, Fig. 1-3. Close to the ½ mile milepost, an anticlinal fold can be seen across the river.

At about this point and across Old Buena Vista Road stands Clifton, the early 19th-century country house built by Major John Alexander. Directly in front of Clifton one of the arms of the former Y-shaped track section exits the Trail; the second exits 225 yards farther, between the two creek bridges. If you walk up this arm to the road you can see the excavation done by the railroad to make this turnaround. From this point the trains backed into Lexington. About 40 yards farther along, the Trail crosses Mill Creek and then passes through the first of the gates that keep the Trail closed to vehicles—and keep the livestock from visiting their neighbors.

Just upstream from the gate are the first red cedar trees you will encounter along the Trail. The cedar's aromatic, rose-brown wood is used for fences, roofing, cedar chests, cabinets, and pencils. Its small, hard berries are consumed by dozens of bird species, including quail, grouse, pheasant, and dove; opossums are also partial to them. The seeds, however, pass through a bird's digestive tract undamaged, which is why the tree spreads so rapidly into untended fields and along fences.

SECTION 3: FROM MILL CREEK TO REID'S DAM

If you have chosen to drive to Mill Creek along Old Buena Vista Road, you'll find a small parking area on your left, outside of a state maintenance yard with a chain link fence. Directly across from it and down a gravel shoulder is the second entrance to the Trail, the Mill Creek entrance, at the Mill Creek gate. If you are entering at this point, pause at the bottom of the gravel shoulder and look west. You'll see the two wooden footbridges that hikers who started at the beginning of the Trail will just have traversed. These cross Mill Creek, where the ruins of old stone walls below the Trail mark the path of the old mill race.

The "Start/Finish" post at Mill Creek gate roughly coincides with old milepost 1, which is about a hundred yards downstream. A small wet-weather spring emerges from the cliff face just downstream from the gate.

Photo by Ed Spencer

*Rapids mark a spot where detritus that washed
out of Mill Creek has settled.*

After passing through the gate, walk down to the river and look back upstream toward Mill Creek. Rocks and soil that washed out of the Mill Creek valley have been deposited in the Maury River channel. When the amount of water in the Maury is low, the river does not have enough power to move the deposits downstream. If you look downstream here, it is clear that the river is flowing in a large meander loop. From turbulence in the water, you can see that the main path of flow passes around the deposits at the mouth of Mill Creek and then crosses the river to the outside of the bend. Over a long period of time, channel fill is removed from the outside of bends and deposited on the inside of the curves. Undercutting of the outside of bends eventually causes unstable cliffs to form. Blocks fall off the cliffs and eventually break up into fragments the river can move downstream. Gradually the erosion causes the valley to become wider. Two such large blocks of rock, each of which contains a drill hole, have fallen from the cliff onto the edge of the Trail. These are located about 750 and 825 feet downstream from the start marker at the Mill Creek entrance gate. Iron stains produced by the chemical alteration of pyrite (an iron sulfide mineral known as "fool's gold") in the limestone are visible on these blocks.

Most of the rock exposed in the cliff, located on your left as you look downstream, is a massive black limestone, part of the Edinburg formation. The bedding can be seen in the face of the cliff. Layers that were deposited horizontally have been tilted to steep angles. Some are nearly vertical. Farther along in this outcrop, the exposed rock contains more shale.

The river flows almost due west along this stretch. Just before the river goes into another turn, you pass an old quarry site from which stone was removed to build the bed of the railroad. Evidence of drill holes for dynamite charges makes it clear how the Trail was blasted out of the cliff face. The bank along the outside of this curve is largely fill, placed during construction of the railroad tracks. Large floods occasionally have stripped the fill from this portion of the Trail; water was about 10 feet above the Trail during the peak of the 1985 flood. The fine sediment seen across the river is deposited in the relatively quiet water on the inside of the curve.

About 400 yards beyond milepost 1, the roadbed was built above the floodplain, which lies about 15 feet below the Trail. The river swings across this floodplain and flows on the south bank against an outcrop of limestone at the base of a cliff on the outside of the curve.

About 200 yards beyond the beginning of the floodplain, you can see across the river that a very large block of rock has broken loose from the cliff face where it has been undercut by the river. Folds that developed in the Edinburg formation are exposed in the bare cliff face. The outcrops there, as all along most of the river, dip to the southeast, as a result of the deformation that accompanied the mountain building and pushed the rock formations in a northwesterly direction.

When the Trail first opened, this stretch was lined with red cedar, staghorn sumac and trumpet-creeper vines, but now you may see more of box elder and autumn olive. Both show the "creep" of invasive species. If you find the sumac, easier to see in the fall, you'll grasp why it's so named, as the shrub's branches resemble a deer's antlers "in velvet." It is a favorite browse of the whitetail deer and the cottontail rabbit. Birds and small mammals such as the skunk feed on the hairy red fruit.

Just upstream from Reid's Dam and Lock, the river cuts across a beautifully developed terrace about 10 feet below the Trail. Much of this terrace is composed of silt that settled in the quiet water behind the dam. This is a common problem with most dams. Eventually silt fills the reservoir behind dams and they lose their water storage capacity. Reid's Dam was breached during the flood of 1969. Water continued to flow through the canal boat locks, however, until the 1985 flood widened the breach in the dam and lowered the riverbed. The river now flows through the lock only during high water. On various occasions plans have been considered for repairing the dam and using it again to generate electrical power. (Note: in this area you'll pass old milepost 5, which marks mile 5 westbound from Buena Vista; and also one of the new short posts, which marks mile 1 from the Mill Creek gate, not from Jordan's Point.)

Photo by Patte Wood

The top of Reid's Lock may safely be examined from behind the railing at a small observation clearing just off the Trail.

Just downstream of Reid's Lock, a well-defined path to the river's edge takes you to a broad rock ledge at waterside, a good spot to rest, picnic, fish or just dangle your feet in the water, or pull a boat over for a break.

The ledge itself is an excellent exposure of the Edinburg formation. Soil washed off this outcrop when the lock was opened. Secondary calcite veins and bedding in the Edinburg are exposed. Many of the structural features seen here formed when the ancient sediment was soft. As a soft mud it moved readily down the slope of the prehistoric sea floor. Bedding becomes flatter at the end of this outcrop until finally a single bed forms a rather large surface. Here, you can see the top of the bedding as well as the edges of the beds. At the downstream end of this outcrop the contact between the solid limestone bedrock and the deposits of gravel, sand, and soil on the flood plain is exposed.

SECTION 4: FROM REID'S DAM TO I-81

At new milepost 1½ is a dense bamboo thicket; its origins are unknown—perhaps it was begun by an escapee from a 19th-century ornamental garden—but it is the progenitor of numerous bamboo stands in Lexington and vicinity. Beyond the bamboo is the first of the Trail's two large stands of horsetails (scouring rush), a leafless, segmented relative of the ferns. The second stand is located along the Trail in the area of the ruins of the South River Lock.

A new geologic map of the Lexington 1:24,000 quadrangle was printed in 2007. This map, done by Gerald Wilkes and published by the Virginia Division of Geology and Mineral Resources, includes the portion of the Chessie Trail from Jordan's Point to several thousand feet downstream from South River. Based on rock exposures off the Trail, a belt of the Martinsburg Formation – the rock unit that lies stratigraphically above the Edinburg Formation – crosses the Trail. The upper contact is shown crossing the Trail close to the western end of a long island in the river. The eastern edge of this belt is defined by a thrust fault that carries the Edinburg onto the Martinsburg. Unfortunately, because the Martinsburg erodes so easily, it is not exposed along this section of Trail.

Just before you pass under Interstate 81, you'll cross Warm Run on a short wooden footbridge. The floodplain here under I-81 is a broad meadow, a good wildlife corridor. You'll notice a wetland just past the I-81 overpass; it may be a remnant of Warm Run's original course, truncated by the highway, or the result of runoff from the highway. A depressed area running north of the Trail and parallel to it at this point suggests that the riverbed may once have been north of where the C&O tracks were laid. A series of hollows along the Trail's north side here occupy the depressed area and provide good wildlife habitat.

Photo by Patte Wood

Hikers cross the wooden footbridge over Warm Run.

SECTION 5: FROM I-81 TO SOUTH RIVER

One of the most important thrust faults in the Central Appalachians crosses the Trail under the bridge for I-81 over the Maury River. This fault is exposed at the southwestern edge of the bridge. Unfortunately it is unlawful to stop a car on the interstate highway, but you might look for it the next time you drive south across the bridge. The fault is not exposed along the Trail, but it is projected to cross the Trail near the first gate east of the I-81 bridge.

Photo by Patte Wood

An I-81 overpass crosses a meadow that overlies one of the most important thrust faults in the Central Appalachians.

Along this fault, known as the Staunton or Staunton-Pulaski fault, Cambrian age rocks of the Elbrook Formation, composed primarily of thin-bedded dolomite with some limestone, have been moved up onto Ordovician age rocks of the Edinburg Formation. Layers of rock nearly a mile thick are missing along the fault, among them entire formations including the Conococheague, Stonehenge, Beekmantown, New Market and Lincolnshire. Rocks on the eastern side of this fault moved great distances to the northwest during the mountain building, carrying a thick slab of rock that includes the Blue Ridge Mountains. The fault lies at depth beneath the Blue Ridge and continues tens, perhaps more than a hundred miles to the southeast. This movement took place more than 200 million years ago when Africa collided with North America, resulting in the formation of the Appalachian Mountains.

As the hiker proceeds downstream from the first gate east of the I-81 bridge, exposures in the far cliff are the

Cambrian age (Fig. 1.1) Elbrook formation, inclined about 30 degrees toward the southeast. The contact between the Elbrook and the overlying Conococheague crosses the Trail about 50 yards upstream of the old BF 16 cement marker. The Elbrook, which is easily eroded, is not exposed, but the contact is near the first part of a long outcrop of the Conococheague limestone and dolomite. A series of riffles show up in the river where resistant layers of rock cross the channel. These same layers are exposed along the Trail. About 200 yards downstream from the BF 16 marker, a steeply dipping fracture, possibly a fault, dips to the northwest.

About 200 feet downstream from the BF 16 post, look up and you can see the tan-colored lower side of the bedding surface that shows good examples of ripple marks and mud cracks formed in the shallow water when the sediment was originally deposited.

Photo by Ed Spencer

*Ripple marks reflect the action of the shallow water
in which the sediment that formed the rock was deposited.*

The rocks exposed all the way from BF 16 to Zimmerman's Lock belong to the Cambrian age Conococheague Formation, which lies above the Elbrook Formation and is about 2,000 feet thick. Much of this formation is a blue limestone with thin layers of silt, but it also includes some layers of dolomite and sandstone. By the time you have reached the next gate, the dips in the cliff are low, and the bedding is inclined at no more than 20 degrees. If you follow the orientation of the bedding as you walk along the Trail you will discover large-scale folds.

The Trail now pulls away from the floodplain and enters a half-mile section having a parkway-like appearance, as it is lined with large box elders and ailanthus that overarch the path.

A hiker traverses the "Boulevard" Section of the Trail.

Photo by Ed Spencer

Just beyond the gate, on the river side, is the foundation of a building mentioned in the Archaeology chapter – possibly a lock keeper's house or a maintenance building for the South River lock. On this stretch of the Trail you are passing through private pasture land. Please keep dogs on leashes and move quietly through.

Photo by Patte Wood

*Calves in a pasture along the "boulevard" stretch are a
reminder to keep to the Trail and not alarm the livestock.*

About midway between the pasture's two gates the
fence jogs down to enclose a spring. At this point the
rocks around the spring are characteristic deep blue
limestones of the Conococheague formation. At 20 yards
downstream from the spring, excellent examples of flat-
pebbled conglomerates are exposed. These structures,
formed when waves broke up thin layers of sediment
over the sea floor, are difficult to see. You will need to
get close to the rock to see them.

On the river side of the Trail you can still see the
remains of a railroad siding that runs up to the ruins
of a 60-foot-long wharf. The metal-roofed building
just beyond the next gate is a spring house originally
constructed at the time the dam was built. At this point
the Trail is close to County Road 703, which was called
"Freehollow Road" because the area was settled by a
number of freed slaves following the Civil War. Across
this road you may see the "Haunted House" ruins.

About 65 yards downstream from the spring and on the road side of the Trail just past the gate, a few chert nodules and stringers are present in the limestone; look for dark, irregular-shaped spots in the lighter Conococheague. In these outcroppings you can also see sand and silt layers in the dolomite bedrock. A few of the chert nodules, which are insoluble to water, protrude from the outcrop. Chert, also known as flint, is a silicon dioxide deposit formed in the sediment on the sea floor. It commonly occurs as nodules or stringers, but in a few places it occurs in beds several feet thick. The origin of the silicon is debated by geologists. One popular explanation is that the silica came from dust blown into the sea, which then settled to the sea floor and recrystallized as chert.

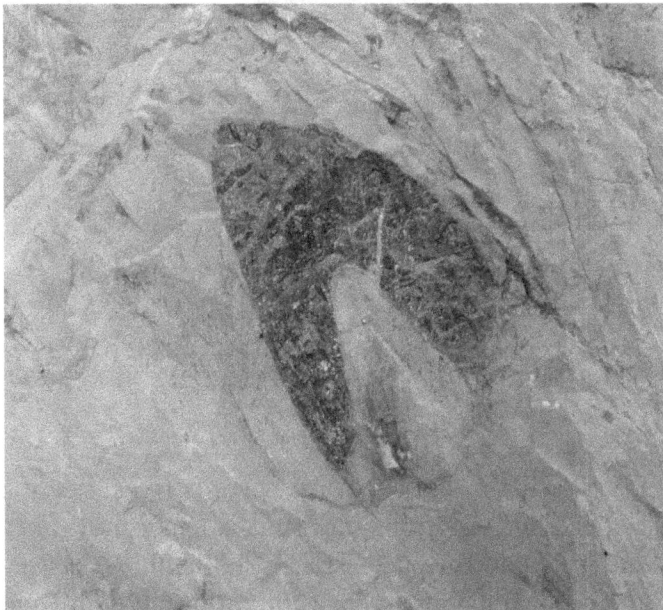

Photo by Ed Spencer

Chert nodules may be seen in the Conococheague just beyond the South River gate.

On the river side just beyond a gate are the remains of the South River Dam. The foundations of a mill were built into what may have been a guard lock for a never-finished canal into the South River valley. Across the river you can see the lock where the canal boats left the river for a half-mile-long canal built to bypass the confluence of the two rivers and to reach the slack water of Ben Salem Dam. The canal has been filled in.

Photo by Patte Wood

A sign just inside the South River gate welcomes
westbound walkers to the central stretch of the Trail.

For kayakers and canoeists, there is river access in this area, but not from the Trail. Canoeists/kayakers should take Old Buena Vista Road to South River Road in the Mountain View area. Turn north on South River Road, and within a few hundred yards you'll find a pullover on the left that allows access to the river. Paddle downstream and you'll soon be on the Maury alongside the Trail. Best destinations for exiting the river are at Ben

Salem Wayside off Route 60 or just above the Robey Bridge in Buena Vista, near the end of the Trail's eastern parking area.

SECTION 6: NAVIGATING THE
SOUTH RIVER JUNCTION

About 300 yards beyond the gate are a parking area, the remains of the footbridge across the South River, and the highway bridge beyond which Stuartsburg Pike (County Road 703/608) makes an abrupt right turn and changes its route number.

The old Trail bridge over South River was washed out during Hurricane Isabel in 2003. At this point, the only way currently to continue on the Trail is to briefly leave the river and cross the South River on the highway bridge. The continuation of Route 703 toward Buena Vista is Route 608. Continue along Route 608 three-tenths of a mile to a junction with Route 839 (Old Sheppard Road), which leads downhill to a gate with a stile on your left. Climb the stile to cross the fence and continue on the Trail across a large field. The Trail here appears to be a dirt road. For those who wish to avoid hiking along the highway, there is a shoulder where parking is possible close to the junction of Routes 608 and 839.

SECTION 7: SOUTH RIVER TO
BEN SALEM LOCK

At South River you have an excellent view of the Blue Ridge, the South River valley, and exposures of alluvium of the South River floodplain. The gravels exposed here and in the river below are composed mainly of quartzites that have come down from the Blue Ridge in the upper part of the South River valley. They resemble Silurian quartzites found upstream in the Maury River. The floodplain here was a major site of prehistoric man's activities in the region, discussed in the Archaeology chapter.

After you traverse the pasture off Route 839 and cross the stream at its eastern end, you will pass for the next mile through an area of considerably more diversified tree species, particularly on the relatively less disturbed hill side of the Trail. Here you will find tulip poplar trees, redbud, slippery elm, honey locust, and hazelnut, among others. The concrete pillar of the gauging station is visible across the river beside the road that runs on the old canal towpath. A short distance farther downstream you can see the twin towers and cable that are the remains of a suspension footbridge destroyed by the 1969 flood.

Photo by Ed Spencer

Sand in the Conococheague Formation is one of the deposits from an ancient, shallow sea.

Many exposures of dark blue limestone with ribbon-like streaks of silt or sand, part of the Conococheague

Formation, crop out along the Trail between BF 14 and the Ben Salem Lock ruins. In this section, you will see excellent examples of small folds that are asymmetric toward the northwest, products of the forces involved in the mountain building that took place more than 200 million years ago. You will also see features in the rock layers that formed when the rocks were being deposited in the ocean nearly 500 million years ago. These include good examples of a flat-pebble conglomerate located about 630 feet downstream from BF 14. These are marked by a small orange-colored square. Conglomerates of this type form waves that break up layers of sediment deposited on the sea floor and are good evidence that the water in this sea was very shallow.

Layers of sandstone and very small cross beds crop out 1,800 feet downstream from BF 14 (about 1,100 feet upstream from the gate near Ben Salem Lock). These are marked by an orange dot painted on the rock. Cross beds form where currents move silt or sand in the shallow water along the coast or in streams.

The absence of rock outcrops along the Trail generally indicates that the rocks beneath the soil in those areas are less resistant to erosion than they are elsewhere.

A number of small solution cavities are present in the exposures of limestone in this section. The entrance to one large cave is seen about 20 feet above the Trail 1,000 feet upstream of the gate near the site of the Ben Salem Lock.

SECTION 8: BEN SALEM LOCK TO ZIMMERMAN'S LOCK RUINS

At the Ben Salem Lock, the conditions that led to the construction of the dam and lock at this particular point are evident. The cliffs on both sides of the river and the narrow floodplain made this an economically attractive spot for construction of a dam. Now only a few stones across the river channel mark where the dam was constructed.

Some of the most interesting structural features along this section of the Trail are exposed in the outcrops at and near the Ben Salem Lock gate. For most of the distance from BF 14 the layers of limestone and dolomite are inclined about 30 degrees to the southeast, but close about 100 feet upstream of the gate the layers show a pronounced curve near ground level. From that point to the gate the layers become steeper until they are nearly vertical at the gate. A few yards downstream from the gate the layers are nearly vertical, but if you look up the slope you can see them change dip and begin to roll over to the east. You are in the core of a large asymmetric anticline. Several small southeast-dipping faults cut and offset the vertically inclined layers of sandstone, silt, dolomite, and limestone a few feet upstream from the gate. A number of small-scale asymmetric folds are evident in these layers.

Photo by Ed Spencer

A number of minor faults such as those seen here may be found along the Trail.

About 125 feet downstream from the gate, the rock layers have a synclinal structure. The solution cavities close to the axis of this syncline suggest that water moved down the limbs of the syncline and then flowed more or less parallel to the axis. About 175 feet east of the gate a small, northwest dipping fault is present. Geologists refer to structures of this type, which dip in the opposite direction from the push that caused the fault, as "backthrusts." The layers east of it are sharply folded, but become much flatter downstream. For the next few hundred feet the layers form several broad open arches and sags. These can be seen by following the stratification in the rocks carefully.

Starting several hundred feet downstream from the gate at Ben Salem lock you get a sweeping view of one of the broadest expanses of the Maury River's floodplain. In 1969 and 1985 flood waters covered US 60 here.

The last large outcrop as you enter the floodplain contains some interesting small scale structural features. The rock consists of thin layers of dolomite interbedded with massive dolomites and limestone. The thin bedded dolomite is weak and yielded to the deformation by forming small "kink-type" folds. The massive layers of dolomite yielded first by stretching accompanied by the formation of cracks now filled with calcite veins, followed by formation of wedge-shaped pieces that slide by one another telescoping the beds. Sausage-shaped bodies called "boudins" formed during the extension. At that time the more plastic (ductile) limestone flowed around the boudins without cracking.

The concrete marker BF 13 is situated at the widest point of the floodplain. Most of the floodplain lies along the inside of this curve. As along other sections of the river, the stream has shifted to the outside of the bend where the cliff has formed. A short distance downstream from BF 13, you will round a bend and catch your first glimpse of what remains of Zimmerman's Lock and Dam. This well-preserved lock always contains water; of the dam only a small portion survives. The cliff from here to the end of the Trail abounds in various members of the oak family. Squirrels and other mammals appreciate their seeds, and in the autumn these trees provide a red backdrop for Zimmerman's Lock.

About halfway between BF 13 and Zimmerman's Lock, look across the river at the low cliff that rises above the stream ahead. Several folds are exposed in this cliff. The limbs of these folds appear to be nearly parallel to one another and nearly horizontal. Such folds are called isoclinal recumbent folds. This appearance may be a result of the angle at which the stream cuts across the axes of the folds.

At the gate by Zimmerman's Lock, notice the beautifully developed small-scale folds defined by thin layers of silt in the Conococheague limestone. About 35 yards east of the gate, a thrust fault zone is exposed. The rock in the fault zone is broken into fragments, called *breccia*. At this place the Elbrook dolomite has been thrust up and over part of the Conococheague Formation. From this point to the end of the Trail, dolomite layers, many of which are thin-bedded platy dolomite of the Elbrook Formation, are prominent in the high cliff along the Trail.

As you approach the Buena Vista end of the Trail, you will see the entrances to a number of caves. Clay that was on the floor of the caves has washed out and has spread down the face of the cliff. The clay remains because it is insoluble, while limestone and dolomites dissolved and were removed by water that moved through the cave. Most caves form below the water table; so, these caves offer additional evidence that over many thousands of years, the river has cut its channel lower until the caves were eventually exposed and the water that once filled them drained out.

Photo by Ed Spencer

The exposed entrance to a cave at the eastern end of the Trail is proof of the river's significant deepening of its channel over time.

About 175 feet from the Buena Vista entrance to the Trail, white calcite veins are present in some layers of dolomite. These veins cut across the beds creating a ladder-like network in some of the more brittle layers. The cracks in which these calcite veins were deposited formed as a result of extension of the brittle dolomite. The calcite was deposited from water that circulated through the cracks when they were below the water table.

You have now reached the eastern entrance to the Trail near the city limits of Buena Vista, seven beautiful and interesting miles from the dedication marker at the Trail head. We hope you had as much fun making the trek as we did and that you will return many times. ✺

A Selected Index of
Names, Places and Events

M

147, 153-4, 164, 172, 177, 186

S

Salem 164, 167, 171
Shenandoah Valley 164
Shenandoah Valley Railroad 164, *See also* Valley Railroad
South River 2-3, 7, 12, 14-15, 18, 27, 124, 140, 145-147, 149-150, 159-161, 169-170, 173, 185-186, 203-204, 209-211
South River Dam 160-161, 184, 186, 210
South River Lock 144, 148, 184-186, 203, 207, 210
Staunton 164, 167, 173, 175
Stono (mansion) 156, 179, 180, 182

T

Tankersley's Tavern 17, 177, 180
Tuscan Villa 186

U

U.S. Geological Survey Gauging Station 12, 15, 212

V

Valley Railroad 164, 167-168, 171, 173,
                See also Shenandoah Valley Railroad
VMI 1, 156, 168, 174, *See Also* Virginia Military Institute
VMI Island 178-179
Virginia Military Institute vii, *See also* VMI

W

Wood, John 180
Woods Creek 2, 7, 16, 178, 192, 194

Z

Zimmerman's Dam 161, 216
Zimmerman's Lock 144, 148-149, 159, 161, 163-164, 187-188, 207, 213, 216
Zimmerman's Lock and Dam 231